SCHOLAR Study Guide
Advanced Higher Chem

Authored by:
Diane Oldershaw (previously Menzieshill High School)

Reviewed by:
Helen McGeer (Firrhill High School)
Nikki Penman (The High School of Glasgow)

Previously authored by:
Arthur A Sandison
Brian T McKerchar
Peter Johnson

Heriot-Watt University
Edinburgh EH14 4AS, United Kingdom.

First published 2019 by Heriot-Watt University.

This edition published in 2019 by Heriot-Watt University SCHOLAR.

Copyright © 2019 SCHOLAR Forum.

Members of the SCHOLAR Forum may reproduce this publication in whole or in part for educational purposes within their establishment providing that no profit accrues at any stage, Any other use of the materials is governed by the general copyright statement that follows.

All rights reserved. No part of this publication may be reproduced, stored in a retrieval system or transmitted in any form or by any means, without written permission from the publisher.

Heriot-Watt University accepts no responsibility or liability whatsoever with regard to the information contained in this study guide.

Distributed by the SCHOLAR Forum.

SCHOLAR Study Guide Advanced Higher Chemistry

Advanced Higher Chemistry Course Code: C813 77

ISBN 978-1-911057-68-0

Print Production and Fulfilment in UK by Print Trail www.printtrail.com

Acknowledgements

Thanks are due to the members of Heriot-Watt University's SCHOLAR team who planned and created these materials, and to the many colleagues who reviewed the content.

We would like to acknowledge the assistance of the education authorities, colleges, teachers and students who contributed to the SCHOLAR programme and who evaluated these materials.

Grateful acknowledgement is made for permission to use the following material in the SCHOLAR programme:

The Scottish Qualifications Authority for permission to use Past Papers assessments.

The Scottish Government for financial support.

The content of this Study Guide is aligned to the Scottish Qualifications Authority (SQA) curriculum.

All brand names, product names, logos and related devices are used for identification purposes only and are trademarks, registered trademarks or service marks of their respective holders.

Contents

1 Inorganic Chemistry — 1

1. Electromagnetic radiation and atomic spectra — 3
2. Atomic orbitals, electronic configurations and the periodic table — 19
3. Transition metals — 55
4. Inorganic Chemistry test — 81

2 Physical Chemistry — 87

1. Chemical equilibrium — 89
2. Reaction feasibility — 153
3. Kinetics — 173
4. Physical chemistry test — 187

3 Researching Chemistry — 191

1. Common chemical apparatus — 193
2. Skills involved in experimental work — 195
3. Stoichiometric calculations — 197
4. Gravimetric analysis — 205
5. Volumetric analysis — 211
6. Practical skills and techniques — 223
7. Researching chemistry test — 239

4 Organic Chemistry and Instrumental Analysis — 243

1. Molecular orbitals — 245
2. Synthesis — 255
3. Stereochemistry — 305
4. Experimental determination of structure — 323
5. Pharmaceutical chemistry — 357
6. Organic chemistry and instrumental analysis test — 367

Glossary	374
Answers to questions and activities	378

Inorganic Chemistry

1	Electromagnetic radiation and atomic spectra	3
	1.1 Electromagnetic radiation	5
	1.2 Spectroscopy	8
	1.3 Using spectra to identify samples	10
	1.4 Energy calculations	13
	1.5 Summary	15
	1.6 Resources	16
	1.7 End of topic test	17
2	Atomic orbitals, electronic configurations and the periodic table	19
	2.1 Spectra, quanta and ionisation	21
	2.2 Quantum numbers	23
	2.3 Atomic orbitals	24
	2.4 Orbital shapes	26
	2.5 Electronic configurations	28
	2.6 Writing orbital box notations	31
	2.7 Ionisation energy	35
	2.8 Covalent bonding	38
	2.9 Dative covalent bonds	40
	2.10 Lewis diagrams	41
	2.11 Shapes of molecules and polyatomic ions	43
	2.12 Examples of molecules with different shapes	43
	2.13 Summary	48
	2.14 Resources	48
	2.15 End of topic test	49
3	Transition metals	55
	3.1 Electronic configuration	57
	3.2 Oxidation states and oxidation numbers	58
	3.3 Ligands and transition metal complexes	61
	3.4 Colour in transition metal complexes	66
	3.5 UV and visible spectroscopy	69
	3.6 Catalysis	74

3.7	Summary	76
3.8	Resources	76
3.9	End of topic test	77

4 Inorganic Chemistry test **81**

Unit 1 Topic 1

Electromagnetic radiation and atomic spectra

Contents

1.1 Electromagnetic radiation	5
1.2 Spectroscopy	8
1.3 Using spectra to identify samples	10
1.4 Energy calculations	13
1.5 Summary	15
1.6 Resources	16
1.7 End of topic test	17

Prerequisites

Before you begin this topic, you should know:

- the electronic structure of atoms (National 4/5 and Higher Chemistry: Chemical changes and structure).

UNIT 1. INORGANIC CHEMISTRY

Learning objective

By the end of this topic, you should know:

- there is a spectrum of electromagnetic radiation;
- electromagnetic radiation can be described in the terms of waves;
- electromagnetic radiation can be characterised in terms of wavelength or frequency;
- the relationship between wavelength and frequency is given by $c = f \times \lambda$;
- the wavelength of visible light is usually expressed in nanometres;
- electromagnetic radiation can be described as a wave and as a particle and is said to have dual nature;
- a photon carries quantised energy proportional to the frequency of the radiation;
- absorption or emission of electromagnetic radiation causes it to behave more like a stream of particles called photons;
- the energy lost or gained by electrons associated with a single photon is given by $E = hf$ or $E = \frac{hc}{\lambda}$;
- it is more convenient for chemists to express the energy for one mole of photons as $E = Lhf$ or $E = \frac{Lhc}{\lambda}$
- energy is often in units of kJ mol^{-1};
- when energy is transferred to atoms, electrons within the atoms may be promoted to higher energy levels;
- atomic emission spectra are made up of lines at discrete frequencies;
- in emission spectroscopy, high temperatures are used to excite the elctrons within atoms - as the electrons drop to lower energy level, photons are emitted;
- an emission spectrum of a sample is produced by measuring the intensity of light emitted at different wavelengths;
- photons of light energy are absorbed by atoms when electrons move from a lower energy level to a higher one;
- an absorption spectrum is produced by measuring how the intensity of absorbed light varies with wavelength;
- each element produces a unique pattern of frequencies of radiation in its emission and absorption spectra;
- atomic emission spectroscopy and atomic absorption spectroscopy are used to identify and quantify the elements present in a sample;
- the concentration of an element within a sample is related to the intensity of light emitted or absorbed.

TOPIC 1. ELECTROMAGNETIC RADIATION AND ATOMIC SPECTRA

1.1 Electromagnetic radiation

Electromagnetic radiation is a form of energy. Light, x-rays, radio signals and microwaves are all forms of electromagnetic radiation. Visible light is only a small part of the range of the **electromagnetic spectrum**.

Figure 1.1: The electromagnetic spectrum

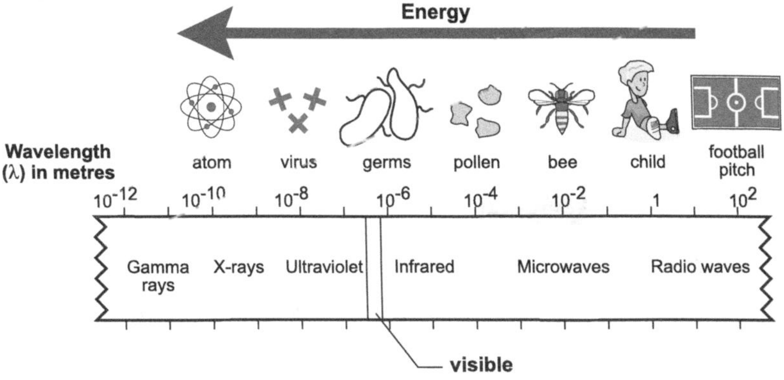

The figure shows that the electromagnetic spectrum has a variety of ways in which it can be described. At the highest energy, the waves are so tightly packed that they are less than an atom's width apart, whilst at the low energy end the waves are a football pitch or greater apart. In the wave model description of electromagnetic radiation, the waves can be specified by their wavelength and frequency. All electromagnetic radiation travels at the **same velocity**. This constant is the speed of light, symbol 'c', and, in a vacuum, it is approximately equal to

$$c = 3.00 \times 10^8 \text{ m s}^{-1}$$

In the following waveform diagram, two waves are travelling with velocity c. In one second, they travel the same distance.

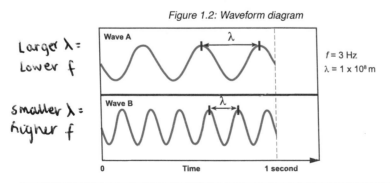

Figure 1.2: Waveform diagram

Larger λ = Lower f

Smaller λ = higher f

Wave A: $f = 3$ Hz, $\lambda = 1 \times 10^8$ m

Formulae are found on page 4 of the SQA CfE Higher and Advanced Higher Chemistry data booklet.

Wavelength has the symbol λ (lambda). It is the distance between adjacent crests (or troughs) and is usually measured in metres or nanometres (1 nm = 10^{-9} m). In the waveform diagram, wave A has a wavelength twice the value of wave B.

Frequency has the symbol f. It is the number of wavelengths that pass a fixed point in one unit of time, usually one second. Frequency is measured as the reciprocal of time (s^{-1}), more commonly called 'hertz' (Hz). In the waveform diagram, wave A has half the frequency value of wave B.

Frequency and wavelength are very simply related. Multiplying one by the other results in a constant value called 'c', the speed of light.

$$c = \text{wavelength} \times \text{frequency}$$
$$c = f \times \lambda$$

The wavelength of wave B (in the waveform diagram) can be checked using this equation.

$$c = \text{speed of light} = 3 \times 10^8 \text{ m s}^{-1}$$
$$f = \text{frequency} = 6 \text{ Hz (or s}^{-1})$$
$$c = f \times \lambda$$
$$\lambda = \frac{c}{f}$$
$$\lambda = \frac{3 \times 10^8 \text{ m s}^{-1}}{6 \text{ s}^{-1}}$$
$$\lambda = 5 \times 10^7 \text{ m}$$

Notice that this is half the value given for wave A (in the waveform diagram).

Since $c = wavelength \times frequency$ and c is a constant, this means that the higher the frequency, the lower the wavelength (and vice versa).

Try Question 1, basing your method on the last example shown. The full working is available in the answer at the end of this booklet and can be used if you have difficulty. You are strongly advised to try these questions on paper. Seeing the solution is not the same as solving the problem!

Obtaining wavelength from frequency　　　　　　　　　　　　　　　Go online

Q1: A beam of light from a sodium street lamp is found to have a frequency of 5.09×10^{14} Hz.
Calculate the wavelength of this light to the nearest nanometre.

....

Q2: Electromagnetic radiation is found to have a frequency of 8.00×10^{12} Hz.
Calculate the wavelength of this radiation. Give your answer to three significant figures.

....

TOPIC 1. ELECTROMAGNETIC RADIATION AND ATOMIC SPECTRA

Q3: Electromagnetic radiation is found to have a frequency of 7.00×10^{14} Hz.
Calculate the wavelength of this radiation. Give your answer to three significant figures.

....

Q4: Electromagnetic radiation is found to have a frequency of 8.00×10^9 Hz.
Calculate the wavelength of this radiation. Give your answer to three significant figures.

....

Q5: A typical microwave oven operates at a frequency of 2.45×10^9 Hz.
Calculate the wavelength of this radiation. Give your answer in centimetres.

In other situations, the wavelength may be given and the value of frequency can be calculated by rearranging the equation to

$$f = \frac{c}{\lambda}$$

Obtaining frequency from wavelength — Go online

Q6: Data book information records the flame colour of potassium as lilac and with a wavelength of 405 nm.
Calculate the frequency of this radiation.

....

Q7: Use the data book to find the wavelength of light emitted by a sample of copper in a flame and thus calculate its frequency. The frequency in hertz is:

a) 5.75×10^{-4}
b) 1.57×10^{20}
c) 5.75×10^{14}
d) 1.74×10^{-15}

....

Q8: Electromagnetic radiation is found to have a wavelength of 1900 nm.
Calculate the frequency of the radiation. Give your answer to three significant figures.

....

Q9: Electromagnetic radiation is found to have a wavelength of 2300 nm.
Calculate the frequency of the radiation. Give your answer to three significant figures.

....

Q10: Electromagnetic radiation is found to have a wavelength of 1500 nm.
Calculate the frequency of the radiation. Give your answer to three significant figures.

© HERIOT-WATT UNIVERSITY

> **Electromagnetic radiation table** Go online
>
> This is a summary table showing the relationship between the symbols for the various radiation characteristics and their units and descriptions.
>
> **Q11:** Complete the table using the terms provided.
>
> | Quantity | Symbol | Unit | Description |
> |---|---|---|---|
> | speed | | | |
> | | λ | | |
> | | | | wave cycles per second |
>
> Terms: c, f, frequency, rate of travel, wave cycles per second, wavecrest separation, wavelength.

1.2 Spectroscopy

distribution of electromagnetic radiation on according to to wavelength & frequency

Visible light is only a small part of the electromagnetic spectrum. A wider range would stretch from gamma rays to radio and TV waves (Figure 1.3).

Figure 1.3: Electromagnetic spectrum

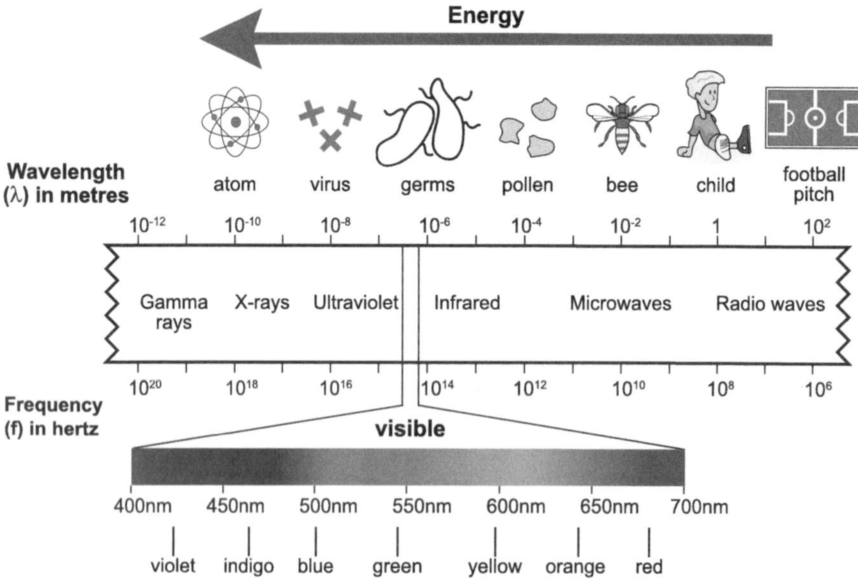

When a beam of white light is passed through a prism or from a diffraction grating onto a screen, a continuous spectrum is seen Figure 1.4 (a). The same effect can be seen in a rainbow.

However, if the light source is supplied by sodium chloride being heated in a Bunsen burner flame, the spectrum turns out not to be a continuous spectrum, but a series of lines of different wavelengths and thus of different colours.

Spectra that show energy being given out by an atom or ion are called atomic emission spectra as shown in Figure 1.4 (b). The pattern of lines in such a spectrum is characteristic of each element and, like a fingerprint, can be used to identify the element.

Figure 1.4: Types of spectrum

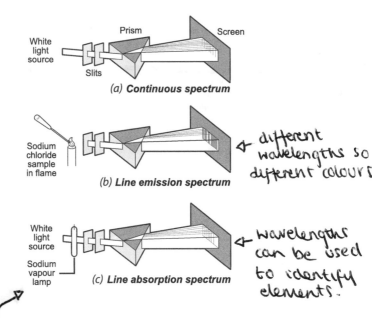

If a beam of continuous radiation like white light is directed through a gaseous sample of an element, the radiation that emerges has certain **wavelengths missing**. This shows up as dark lines on a continuous spectrum and is called an atomic absorption spectrum, see Figure 1.4 (c).

This also provides a pattern that can often be used in identification. In both techniques some lines normally occur in the visible region (400-700 nm) but some applications use the ultraviolet region (200-400 nm). Both emission and absorption spectroscopy can be used to determine whether a certain species is present in a sample and how much of it is present, since the intensity of transmitted or absorbed radiation can be measured.

In atomic absorption spectra electromagnetic radiation is directed at an atomised sample. The electrons are promoted to higher energy levels as the radiation is absorbed. The absorption spectrum is a measure of the sample's transmission of light at various wavelengths.

In atomic emission spectra high temperatures are used to excite electrons within the atoms. As they fall back down to their original energy level photons are emitted. The emission spectrum is a measure of the light emitted by the sample at different wavelengths.

Absorption spectrometer   Go online

An online and printable electromagnetic spectrum is available from the Royal Society of Chemistry.

View the videos on spectroscopy on the Royal Society of Chemistry's website.
http://www.rsc.org/learn-chemistry/resource/res00001041/spectroscopy-videos

---

1.3 Using spectra to identify samples

Spectra can be used to give information about how much of a species is present in a sample. For example the concentration of lead in drinking water or a foodstuff can be found. First a calibration graph is prepared from known concentrations of lead solutions. The radiation absorbed by these samples is plotted against concentration and when the unknown sample is analysed the concentration of lead can be found from the graph.

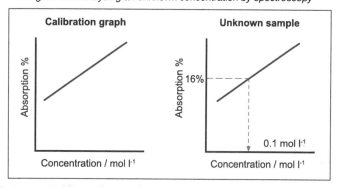

Figure 1.5: Analysing an unknown concentration by spectroscopy

The figure shows the build up of a calibration graph on the left as the radiation absorbed is measured at different concentrations. After the graph is complete, the unknown sample is measured (in this case at 16% absorption) and reading off from the graph shows a concentration in the sample of 0.1 mol l^{-1}.

Using spectra to identify samples

Go online

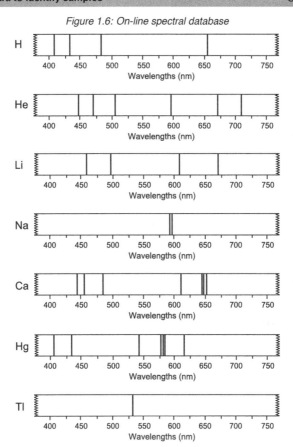

Figure 1.6: On-line spectral database

In this activity you can use the online database or the printed version to answer the questions about these two spectral problems.

Part a) This spectrum was obtained from the atmosphere around the Sun.

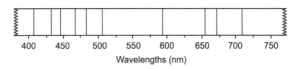

Q12: Write the name of the element which is responsible for the line at 655 nm.

....

Q13: Write the name of the other element present.

Part b) These were obtained from equal sized soil samples that have been treated and the spectra measured.

1. **Sample A.** From productive farmland
2. **Sample B.** From the site of an old factory site where insecticides and pesticides were produced.

Sample A

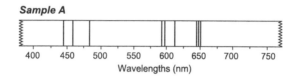

Sample B

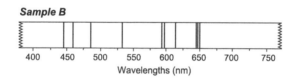

The boxes in this grid contain elements whose spectra appear in the database.

A	B	C
Li	Hg	He
D	**E**	**F**
Tl	Na	Ca

Q14: Which element is responsible for the triplet of lines around 650 nm?

....

Q15: Which other metal is present in *both* samples?

....

Q16: Use the database to work out which element is present in sample (B) but not in (A).

....

Q17: Mercury is a metal whose salts are well known poisons and thallium salts are used in some countries as a rat poison. Is there evidence of mercury in sample (A)?

....

Q18: Is there evidence of mercury in sample (B)?
....

Q19: Write the *name* of the element which you think the spectra prove to be the cause of the pollution.

1.4 Energy calculations

The full emission spectrum for the hydrogen atom consists of a number of series of lines, named after the scientists who first investigated the spectra. Only the Balmer series lies in the visible region.

Figure 1.7: Hydrogen spectrum

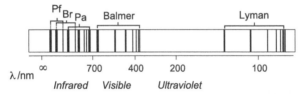

(Pf = Pfund Br = Brackett Pa = Paschen Br overlaps Pf and Pa)

The spectral lines of radiation emitted by the hydrogen atom in the spectrum (Figure 1.7) show emission at only certain frequencies. Since electromagnetic radiation carries energy related to the frequency, only certain precise energy values are being involved.

Electromagnetic radiation is often described as having a dual nature. Max Planck developed the theory that as well as electromagnetic radiation being described by a wave, under certain circumstances it can be regarded as a stream of particles. These particles are called photons.

The energy carried by a photon is related to its frequency by the equation:

$$E = hf$$

where h is **Planck's constant** = 6.63 × 10^{-34} J s

So when dealing with one mole of photons the energy involved would be:

$$E = Lhf$$

where L is **Avogadro's constant** = 6.02 × 10^{23} mol^{-1}

but $c = f \times \lambda$

so $f = \dfrac{c}{\lambda}$

and $E = \dfrac{Lhc}{\lambda}$

Each line in a spectrum has a precise frequency that corresponds to a fixed value of energy. Calculations using these formulae are common.

Example : Obtaining energy values

The red line in the hydrogen spectrum has a wavelength of 656 nm. Calculate:

1. The energy value of one photon of light at this wavelength.
2. The energy value in kJ mol^{-1} for one mole of photons at this wavelength.

$$c = f \times \lambda$$

therefore $f = \dfrac{c}{\lambda}$

and since $E = hf$ for one photon

then $E = \dfrac{hc}{\lambda}$

$$E = \dfrac{6.63 \times 10^{-34} \text{ J s} \times 3 \times 10^8 \text{ m s}^{-1}}{656 \times 10^{-9} \text{ m}}$$

$E = 3.03 \times 10^{-19}$ J (this is for one photon)

So for one mole of photons

$E = 3.03 \times 10^{-19}$ J $\times 6.02 \times 10^{23}$ mol^{-1}

$E = 18.24 \times 10^4$ J mol^{-1}

$E = 1.82 \times 10^5$ J mol^{-1}

$E = 182$ kJ mol^{-1}

Obtaining wavelength values

Q20: Chlorinated hydrocarbon molecules contain **C-Cl** bonds. The energy required to break these is 338 kJ mol^{-1}.

1. Calculate the wavelength of light required to break one mole of these bonds.
2. By reference to the electromagnetic spectrum, suggest why these molecules are unstable in the upper atmosphere.

Energy from wavelength

Go online

Q21: The line in an emission spectrum has a wavelength of 1600 nm.
Calculate the energy value for one mole of photons at this wavelength. Give your answer to three significant figures.

....

Q22: The line in an emission spectrum has a wavelength of 1300 nm.
Calculate the energy value for one mole of photons at this wavelength. Give your answer to three significant figures.

....

Q23: The line in an emission spectrum has a wavelength of 1400 nm.
Calculate the energy value for one mole of photons at this wavelength. Give your answer to three significant figures.

....

Q24: The line in an emission spectrum has a wavelength of 1900 nm.
Calculate the energy value in kJ mol^{-1} for one mole of photons at this wavelength. Give your answer to three significant figures.

1.5 Summary

Summary

You should now be able to state that:

- there is a spectrum of electromagnetic radiation;
- electromagnetic radiation can be described in the terms of waves;
- electromagnetic radiation can be characterised in terms of wavelength or frequency;
- the relationship between wavelength and frequency is given by $c = f \times \lambda$;
- the wavelength of visible light is usually expressed in nanometres;
- electromagnetic radiation can be described as a wave and as a particle and is said to have dual nature;
- a photon carries quantised energy proportional to the frequency of the radiation;
- absorption or emission of electromagnetic radiation causes it to behave more like a stream of particles called photons;
- the energy lost or gained by electrons associated with a single photon is given by $E = hf$ or $E = \frac{hc}{\lambda}$;

> **Summary continued**
> - it is more convenient for chemists to express the energy for one mole of photons as $E = Lhf$ or $E = \frac{Lhc}{\lambda}$
> - energy is often in units of kJ mol^{-1};
> - when energy is transferred to atoms, electrons within the atoms may be promoted to higher energy levels;
> - atomic emission spectra are made up of lines at discrete frequencies;
> - in emission spectroscopy, high temperatures are used to excite the elctrons within atoms - as the electrons drop to lower energy level, photons are emitted;
> - an emission spectrum of a sample is produced by measuring the intensity of light emitted at different wavelengths;
> - photons of light energy are absorbed by atoms when electrons move from a lower energy level to a higher one;
> - an absorption spectrum is produced by measuring how the intensity of absorbed light varies with wavelength;
> - each element produces a unique pattern of frequencies of radiation in its emission and absorption spectra;
> - atomic emission spectroscopy and atomic absorption spectroscopy are used to identify and quantify the elements present in a sample;
> - the concentration of an element within a sample is related to the intensity of light emitted or absorbed.

1.6 Resources

- Royal Society of Chemistry (http://www.rsc.org)
- SSERC (https://bit.ly/2Kw16IL)
- Nuffield Foundation (http://bit.ly/2a79bNI). - spectra formed by gratings.
- Glow Scotland (http://bit.ly/2r8LLPl) (from 'Chemistry: A Practical Guide Support Materials').

TOPIC 1. ELECTROMAGNETIC RADIATION AND ATOMIC SPECTRA

1.7 End of topic test

End of Topic 1 test — Go online

The end of topic test for *Electromagnetic radiation and atomic spectra*

Q25: Which of these has the highest frequency?

a) γ - radiation
b) α - radiation
c) Visible light
d) Radio waves

....

Q26: Electromagnetic radiation may be regarded as a stream of:

a) x-rays.
b) α-particles.
c) electrons.
d) photons.

....

Q27: Which of these happens as the frequency of an orange laser light is decreased?

a) Wavelength goes down.
b) Colour moves towards red.
c) Velocity goes down.
d) Energy increases.

....

Q28: Compared to infrared radiation, ultraviolet radiation has:

a) lower frequency.
b) lower velocity.
c) higher energy.
d) higher wavelength.

....

Q29: Which of the following is used to represent the speed of light?

a) f
b) λ
c) c
d) E
e) h

Q30: Which of the following is used to represent a constant? Choose two.

a) f
b) λ

c) c
d) E
e) h

....

Q31: Which of the following is measured in Hertz?

a) f
b) λ
c) c
d) E
e) h

Look at this emission spectrum for hydrogen.

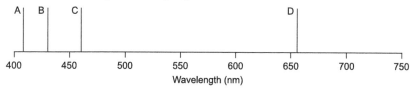

Q32: Which of the lines has the lowest energy (A, B, C, or D)?

....

Q33: Calculate the wavelength, to the nearest nanometre, of the line with a frequency of 7.35 x 10^{14} Hz.

The spectrum of thallium shown has only one line at 535 nm.

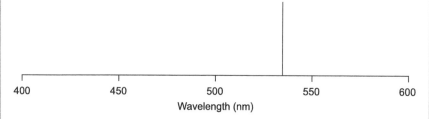

Q34: What colour would thallium salts show up as in a Bunsen flame?

....

Q35: Calculate the energy in kJ mol^{-1} of the emission line that occurs at 535 nm. Give your answer to three significant figures.

Unit 1 Topic 2

Atomic orbitals, electronic configurations and the periodic table

Contents

2.1	Spectra, quanta and ionisation	21
2.2	Quantum numbers	23
2.3	Atomic orbitals	24
2.4	Orbital shapes	26
2.5	Electronic configurations	28
2.6	Writing orbital box notations	31
	2.6.1 The periodic table	34
2.7	Ionisation energy	35
2.8	Covalent bonding	38
2.9	Dative covalent bonds	40
2.10	Lewis diagrams	41
2.11	Shapes of molecules and polyatomic ions	43
2.12	Examples of molecules with different shapes	43
2.13	Summary	48
2.14	Resources	48
2.15	End of topic test	49

Prerequisites

Before you begin this topic, you should understand:

- the structure of the atom in terms of the subatomic particles (National 4 and National 5: Chemical Changes and Structure);
- the electromagnetic spectrum, emission and absorption spectroscopy (Advanced Higher Chemistry);
- how to solve problems and calculations associated with the electromagnetic spectrum, emission and absorption spectroscopy (Advanced Higher Chemistry);
- ionisation energy (Higher Chemistry: Chemical Structures and Changes);
- types of bonding including metallic, non-polar covalent (pure covalent), polar covalent

UNIT 1. INORGANIC CHEMISTRY

> **Prerequisites continued**
>
> and ionic bonding. (Higher Chemistry: Chemical Changes and Structure) (Metallic bonding is not considered in this topic).
>
> - that ionic and polar covalent bonding does not exist in elements. (National 5 and Higher Chemistry: Chemical Changes and Structure).
>
> - electronegativity (attraction of an element for bonding electrons) (Higher Chemistry: Chemical Changes and Structure) values are found in the SQA data booklet.
>
> - that the difference in electronegativity values of elements gives an indication to the likely type of bonding between atoms of different elements. (Higher Chemistry: Chemical Changes and Structure).

> **Learning objective**
>
> By the end of this topic, you should be able to:
>
> - relate spectral evidence to electron movements and ionisation energy;
>
> - describe the four quantum numbers and relate these to atomic orbitals, their shape and relative energies;
>
> - describe the four s, p and d atomic orbitals, their shapes and relative energies;
>
> - relate the ionisation energies of elements to their electronic configuration, and therefore to their position in the periodic table;
>
> - describe the electronic configuration of atoms 1-20 in spectroscopic notation;
>
> - write electronic configurations using spectroscopic notation;
>
> - explain that covalent bonding involves the sharing of electrons and can describe this through the use of Lewis electron dot diagrams;
>
> - predict the shape of molecules and polyatomic ions through consideration of bonding pairs and non-bonding pairs and the repulsion between them;
>
> - understand the decreasing strength of the degree of repulsion from lone-pair/lone-pair to non-bonding/bonding pair to bonding pair/bonding pair.

TOPIC 2. ATOMIC ORBITALS, ELECTRONIC CONFIGURATIONS AND THE PERIODIC TABLE

2.1 Spectra, quanta and ionisation

Light can interact with atoms and provide valuable information about the quantity and type of atom present. Electromagnetic radiation has provided important clues about the actual structure of the atom and the organisational relationship between the elements in the periodic table.

The spectral lines in an atomic emission or absorption spectrum occur at precise frequencies. Since the frequency is related to energy it is obvious that only certain precise energy values are involved (see Figure 2.1).

Figure 2.1: Hydrogen atom spectrum

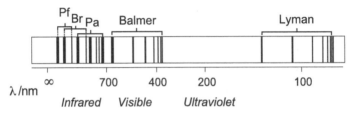

(Pf = Pfund Br = Brackett Pa = Paschen Br overlaps Pf and Pa)

Atomic spectra are caused by electrons moving between different energy levels. These are fixed for any one atom. We say that the energy of electrons in atoms is quantised. Quantum theory states that matter can only emit or absorb energy in small fixed amounts. When an electron in an atom absorbs a photon of energy, it moves from a lower energy level to a higher energy level. When the electron drops back down, energy is emitted (see Figure 2.2).

Figure 2.2: Emission of energy

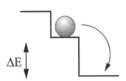
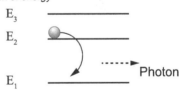

The ball has certain stable levels. Energy is emitted if it falls down.

The electron has discrete energies. Energy is emitted if it falls down.

The energy of the photon emitted is

$$\Delta E = E_2 - E_1 = hf$$

The frequency of the line in the emission spectrum represents the difference in energy between the levels. We call these energy levels shells or sub-shells and in Figure 2.3 the letter n = 1, 2, 3, etc. defines what is known as the principal quantum number.

22 UNIT 1. INORGANIC CHEMISTRY

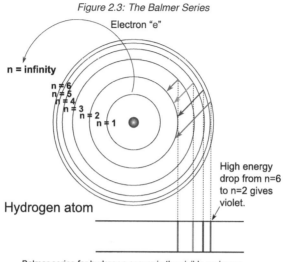

Figure 2.3: The Balmer Series

Balmer series for hydrogen occurs in the visible region as electrons emit photons on transitions down to level **n = 2**.

Figure 2.3 (the Balmer series for hydrogen) shows the spectrum produced when electrons, having been excited into higher energy levels, drop back to the $n = 2$ level and emit radiation in the visible region. Electrons dropping to $n = 1$ level would emit in the ultraviolet (Lyman series). Notice also that the levels get closer together as n increases.

Level number $n = 1$ is the lowest energy level and an electron in a hydrogen atom would occupy this level in the **ground state** under normal conditions.

In Figure 2.3, an electron labelled as electron 'e' is shown **escaping** from level $n = 1$ to infinity. This corresponds to the electron breaking away from the atom completely and represents the **ionisation energy** of that element.

It can be represented in general terms as:

$$X_{(g)} \rightarrow X^+_{(g)} + e^-$$

In the hydrogen atom, the highest energy line in the Lyman series where the lines converge (see Figure 2.1) occurs at a wavelength of 91.2 nm. The ionisation energy can be calculated from this wavelength.

TOPIC 2.   ATOMIC ORBITALS, ELECTRONIC CONFIGURATIONS AND THE PERIODIC TABLE

Calculating ionisation energy

Q1: Calculate the ionisation energy in kJ mol^{-1} for the hydrogen atom from the spectral information that the Lyman series converges at 91.2 nm.

Try solving this problem for yourself using the equations from Topic 1.

2.2 Quantum numbers

The lines in the spectrum of hydrogen are adequately explained by the picture of the atom in the last section. Emission spectra of elements with more than one electron provide evidence of sub-levels within each principal energy level above the first. Quantum theory now defines the allowed energy levels of electrons by four quantum numbers. No two electrons in an atom can have the same four quantum numbers. These quantum numbers can be thought of as 'addresses' for electrons.

For example: Mr Smith lives at Flat 2, number 8, Queen Street, Perth.

His Quantum address	Perth	Queen Street	Number 8	Flat 2
This defines his position	Town	Street	Number	House

If the same framework is considered for Quantum numbers and electrons.

For an electron in an atom	Principal Quantum number	Second Quantum number	Third Quantum number	Fourth Quantum number
This defines the position	Shell	Sub-shell	Direction	Spin

For an electron in an atom each quantum number requires a bit more explanation.

Principal quantum number, symbol n, determines the main energy level. It can have values $n =$ 1, 2, 3, 4, etc. The numbers determine the size and energy of the shell.

Angular momentum quantum number, symbol ℓ, determines the shape of the sub-shell and is labelled as *s, p, d, f*. This can have values from zero to (n -1). The second quantum number is also known as the angular momentum quantum number.

So if n has a value = 4, then ℓ can take these values:

- value 0 labelled as '*s*' subshell;
- value 1 labelled as '*p*' subshell;
- value 2 labelled as '*d*' subshell;
- value 3 labelled as '*f*' subshell.

So for $n = 4$ there are 4 possible subshells. The use of letters (*s, p, d, f*), instead of numbers, aids identification of the sub-shells.

The relationship between n and ℓ is shown in Table 2.1. A useful memory aid is that there are as many subshells as the value of n.

Table 2.1: Relating n and ℓ

Principal Quantum number value n	Second Quantum number value: ℓ	Sub-shell name
1	0	1s
2	0	2s
	1	2p
3	0	3s
	1	3p
	2	3d

The third and fourth quantum numbers will be explained in the next section.

2.3 Atomic orbitals

Before determining where an electron is within a sub-shell and considering the third quantum number, it must be understood that electrons display the properties of both particles and waves. If treated as particles, **Heisenberg's uncertainty principle** states that it is impossible to state precisely the position and the momentum of an electron at the same instant. If treated as a wave, the movement of an electron round the nucleus can be described mathematically. From solutions to these wave equations it is possible to produce a statistical picture of the probability of finding electrons within a given region. Regions of high probability are called atomic orbitals.

Figure 2.4: Shapes of s orbitals

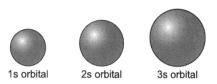

1s orbital   2s orbital   3s orbital

TOPIC 2. ATOMIC ORBITALS, ELECTRONIC CONFIGURATIONS AND THE PERIODIC TABLE

An atomic orbital is the volume in space where the probability of finding an electron is more than 90%. So in Figure 2.4 the s-orbitals shown are spherical in shape, the diameter of the sphere increasing as the value of n increases. At any instant in time there is approximately a 90% chance of finding the electron within the sphere.

Magnetic quantum number (also known as magnetic quantum number), symbol m_ℓ, relates to the orientation in space of the orbital. It is dependent on ℓ because m_ℓ can take on any whole number value between $-\ell$ and $+\ell$.

So if $\ell = 2$ (labelled as a 'd' orbital)

m_ℓ could have the value +2, +1, 0, -1, -2

So for $\ell = 2$ there are five atomic orbitals.

The relationship between ℓ and m_ℓ is shown in Table 2.2

Table 2.2: Relating ℓ and m_ℓ

Sub-shell name	Possible value: ℓ	Possible value: m_ℓ
1s	0	0
2s	0	0
2p	1	-1, 0, +1
3s	0	0
3p	1	-1, 0, +1
3d	2	-2, -1, 0, +1, +2

The relationship between n, ℓ and m_ℓ can be summarised in the next activity.

© HERIOT-WATT UNIVERSITY

26 UNIT 1. INORGANIC CHEMISTRY

Relating quantum numbers   Go online

Q2: Complete the table using the numbers provided.

Value of n	Value of l	Value of m	Subshell name
1	0	0	
2	0		2s
		-1 0 +1	2p
	0	0	3s
	1		3p
	2	-2 -1 0 +1 +2	
-1 0 +1	0	3d	1
1s	3		

2.4 Orbital shapes

Every atomic orbital can hold a maximum of two electrons and has its own shape, dictated by the quantum numbers.

s orbitals: are spherical with size and energy increasing as the value of *n* increases (see Figure 2.4).

p orbitals: have a value of $\ell = 1$ and there are, therefore, three possible orientations in space, corresponding to m_ℓ = -1, 0, +1. The three *p*-orbitals are **degenerate** (have the same energy as each other), and have the same shape, approxiately dumbbell shaped and at right angles to one another (Figure 2.5). Each orbital is defined as if it lies along a set of x, y, z axes. The 2*p* orbitals are thus 2p_x, 2p_y and 2p_z. The 3*p* orbitals would be the same shape but larger and at higher energy.

Figure 2.5: Shapes of p orbitals

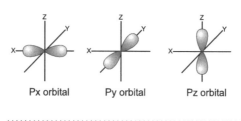

Px orbital   Py orbital   Pz orbital

TOPIC 2. ATOMIC ORBITALS, ELECTRONIC CONFIGURATIONS AND THE PERIODIC TABLE

d orbitals: occur in five different orientations (Figure 2.6) corresponding to the third quantum number m_ℓ = +2, +1, 0, -1, -2 and have labels which come from the complex mathematics of quantum mechanics. These orbitals are important in examination of the properties of transition metals. Note that d orbitals are degenerate (of equal energy) as well as p orbitals in an isolated atom. Three of the d orbitals lie between axes and two d orbitals lie along axes.

Figure 2.6: Shapes of d orbitals

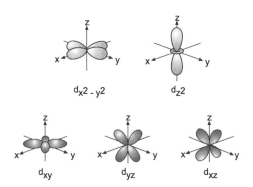

f orbitals: occur in seven different orientations relating to ℓ = 3. The shapes are complex and do not concern us.

Each atomic orbital can hold a maximum of two electrons. Each electron in an orbital has a spin which causes it to behave like a tiny magnet. It can spin clockwise, represented as ↑, or anti-clockwise, represented as ↓. In any orbital containing two electrons they must be paired, with the spins opposed, sometimes represented as ↑↓. When spin representations are shown in a box, the box represents the atomic orbital.

Figure 2.7: Electron spin

spin = -1/2   spin = +1/2

Spin magnetic quantum number: determines the direction of spin. It is therefore called the spin quantum number, m_s. It has values of $+\frac{1}{2}$ or $-\frac{1}{2}$ (See Figure 2.7)

© HERIOT-WATT UNIVERSITY

2.5 Electronic configurations

The arrangement of electrons in the energy levels and orbitals of an atom is called the electronic configuration. This can be expressed in three different ways.

1. Using quantum numbers

The four quantum numbers provide an address for the electron. No two electrons can have the same four quantum numbers, e.g. the three electrons in a lithium atom would have these addresses :

	Principal: n	Second: ℓ	Third: m_ℓ	Fourth: m_s
First electron	1	0	0	$+1/2$
Second electron	1	0	0	$-1/2$
Third electron	2	0	0	$+1/2$

2. Orbital box notation

Each orbital in an atom is represented by a box and each electron by an arrow. The boxes are filled in order of increasing energy. The orbital boxes for the first two energy levels are shown in Figure 2.8.

Figure 2.8: Orbital boxes

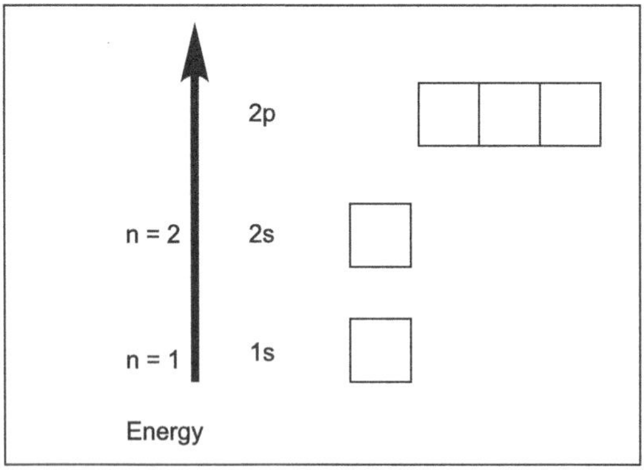

Using this notation, the electronic configuration for lithium (3 electrons) and carbon (6 electrons) is shown (Figure 2.9). Notice the presence of unpaired electrons.

TOPIC 2. ATOMIC ORBITALS, ELECTRONIC CONFIGURATIONS AND THE PERIODIC TABLE

Figure 2.9: Electron configuration of lithium and carbon

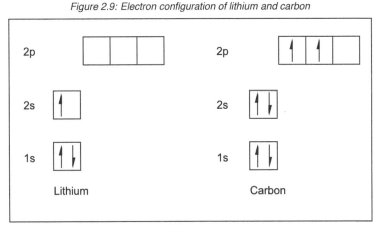

Note that you will sometimes come across full-headed arrows in questions. You can draw either half-headed or full-headed arrows when showing box notation.

3. Spectroscopic notation

This uses a shorthand representation of the arrangement of the electrons, so that the notation taken from Figure 2.9 for carbon would be:

$1s^2 2s^2 2p^2$

Notice that the designation $2p^2$ does not show which of the degenerate orbitals the electrons occupy.

Spectroscopic notations are sometimes made shorter by labelling the core of filled inner shells with the configuration of the preceding noble gas.

For example:

- Neon is $1s^2 2s^2 2p^6$
- Sodium is $1s^2 2s^2 2p^6 3s^1$
- Sodium can be written as $[Ne]3s^1$

Spectroscopic notation

Go online

Q3: Which of these is the spectroscopic notation for a lithium atom?

a) $1s^2\ 2s^2$
b) $2s^1\ 1s^2$
c) $1s^2\ 2s^1$
d) $2s^2 1s^1$

....

Q4: Which element has atoms with the spectroscopic designation [Ar]$4s^1$?

a) Hydrogen
b) Lithium
c) Chlorine
d) Potassium

....

Q5: How many electrons are there in the $2p$ sub-shell of the oxygen atom?

....

Q6: Which number would complete this spectroscopic notation for a nitrogen atom?

$1s^2\ 2s^2\ 2p^?$

....

Q7: Carbon has two **unpaired** electrons. How many **unpaired** electrons would boron have?

....

Q8: Which element is represented as [Ne] $3s^2$?

a) Helium
b) Oxygen
c) Neon
d) Magnesium

....

Q9: Which of these represents the spectroscopic notation of a lithium **ion**? (Remember that a lithium ion has lost an electron to become Li$^+$)

a) $1s^1$
b) $1s^2$
c) $1s^2\ 2s^1$
d) $1s^2\ 2s^2$

2.6 Writing orbital box notations

Multi-electron atoms can be represented in an orbital box diagram by applying three rules to determine the **ground state** electronic configuration.

Rule 1 'AUFBAU'	The **aufbau principle** (from German 'building up') states that when electrons are placed into orbitals, the energy levels are filled up in order of increasing energy, e.g. in Figure 2.9 the $1s$ has filled before $2s$.
Rule 2 'PAULI'	The **Pauli exclusion principle** states that no two electrons in the one atom can have the same set of four quantum numbers. As a consequence, no orbital can hold more than two electrons and the two electrons must have opposite spins, e.g. ↑↑ or ↓↓ are **NOT** allowed.
Rule 3 'HUND'	**Hund's rule** states that when there are *degenerate* orbitals in a sub shell (as in $2p$), electrons fill each one singly with spins parallel before pairing occurs. Thus carbon (Figure 2.9) has not paired up the two $2p$ electrons.

The relative energies corresponding to each orbital can be represented in order as far as $8s$:

Figure 2.10: Aufbau diagram

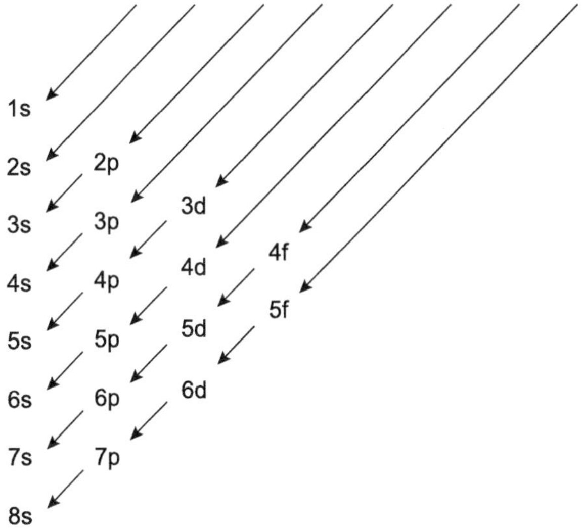

1s
2s 2p
3s 3p 3d
4s 4p 4d 4f
5s 5p 5d 5f
6s 6p 6d
7s 7p
8s

Following the arrows gives the order of orbital filling.

In orbital box notation:

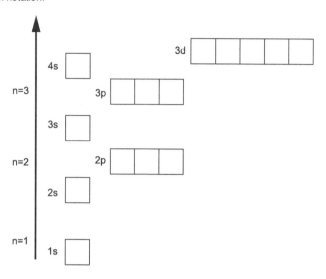

You may have noticed that the 4s orbital has been placed below the 3d in energy and think this is a mistake! It's no mistake, however. Although it is further from the nucleus in terms of space, it is lower in energy and gets filled up first.

Orbital box diagrams and spectroscopic notations can be worked out using the three rules.

Example : Working out the electronic configuration for an oxygen atom.

Rule 1	Start at the 1s level. Place electron 1 (of 8).
Rule 2	Pair up the second electron in 1s. Oppose spin. Repeat for 2s orbital. Four electrons are now placed.
Rule 3	Place electrons 5, 6, 7 into three degenerate 2p. Place electron 8 paired into any 2p.

Spectroscopic: $1s^2 2s^2 2p^4$

TOPIC 2. ATOMIC ORBITALS, ELECTRONIC CONFIGURATIONS AND THE PERIODIC TABLE

Orbital box:

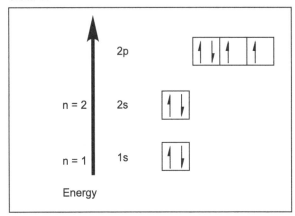

Orbital box notation   Go online

The purpose of this activity is to be able to apply the rules and principles which predict electronic configuration to derive orbital box notation pictures for different elements. This activity allows you to practise working out the orbital box notations for some main group elements in the periodic table.

Q10: Apply the rules and principles which predict electronic configuration to derive orbital box notations for elements 1-20 in the periodic table using the following orbital box diagram.

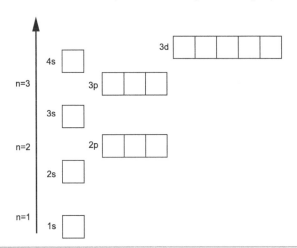

© HERIOT-WATT UNIVERSITY

2.6.1 The periodic table

The structure of the periodic table depends upon the electronic configuration of the elements. Since the chemical properties of an element are dictated by the electrons in the outer shell, the periodic table relates configuration to properties. For example, fluorine has a configuration $1s^2\ 2s^2\ 2p^5$ and thus behaves like other Group Seven elements and is reactive because of its almost complete p sub-shell. The groups and periods of the periodic table are further organised into blocks. Fluorine is thus a p-block element, since the last sub-shell being filled is a p sub-shell.

Periodic table blocks   Go online

The purpose of this activity is to relate the structure of the periodic table to the electronic configuration of the elements.

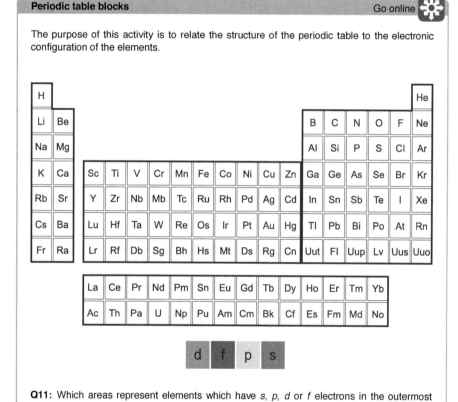

Q11: Which areas represent elements which have s, p, d or f electrons in the outermost sub-shell?

....

Q12: Which block contains the noble gases (excluding Helium)?

a) s
b) p
c) d
d) f

TOPIC 2. ATOMIC ORBITALS, ELECTRONIC CONFIGURATIONS AND THE PERIODIC TABLE

Q13: Which block contains the most reactive metals?

a) s
b) p
c) d
d) f

Q14: What **one word** name is given to the *d*-block elements?

In each of the next four questions classify the element as being *s*, *p*, *d*, or *f* block.

Q15: Aluminium

Q16: $1s^2\ 2s^2\ 2p^6$

Q17: Scandium

Q18: $1s^2 2s^2\ 2p^6\ 3s^1$

2.7 Ionisation energy

The **ionisation energy** of an element is the energy required to remove one mole of electrons from one mole of the gaseous atoms. The second and subsequent ionisation energies refer to removal of further moles of electrons.

It can be represented as:

First Ionisation Energy: $X(g) \rightarrow X^+(g) + e^-$

Second Ionisation Energy: $X^+(g) \rightarrow X^{2+}(g) + e^-$

The variation in first ionisation energies for the first 36 elements (Figure 2.11) relates to the stability of the electronic configuration.

UNIT 1. INORGANIC CHEMISTRY

Figure 2.11: First ionisation energies of the elements

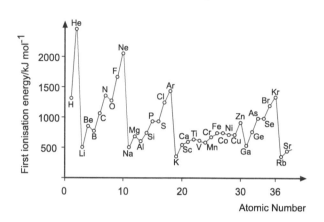

Refer to Figure 2.11 as you consider this question and answer set. Think it through.

Question	Answer
1. Which element group appears at the four peaks?	These are group **0**, noble gases.
2. Would the ionisation be easy or difficult?	**Difficult** as it is a high value.
3. Are these elements reactive or unreactive?	**Unreactive** noble gases.
4. How many electrons in the outer shell?	All are full, usually with **8** electrons.

Conclusion: the ionisation energy evidence supports the electronic configuration theory.

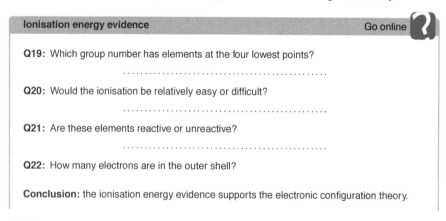

TOPIC 2. ATOMIC ORBITALS, ELECTRONIC CONFIGURATIONS AND THE PERIODIC TABLE

There are two further features of Figure 2.11 which deserve some attention.

Feature 1: There is a slight dip from beryllium to boron.

Beryllium is: $1s^2 2s^2$ and boron is: $1s^2 2s^2 2p^1$

Beryllium has a full sub-shell and is more stable. Boron has a single 2p electron and is less stable.

Feature 2: There is a dip in the middle of the p-block from nitrogen to oxygen, see Figure 2.11.

Nitrogen is: $1s^2 2s^2 2p^3$ and oxygen is: $1s^2 2s^2 2p^4$

In orbital box notation:

Half filled sub-shells are relatively stable and it is easier to remove the fourth **'p'** electron from the 2p shell of oxygen.

Look at Figure 2.11 again and find another part of the graph that can be explained in the same way as features **1** and **2**.

The first, second and successive ionisation energies also provide evidence of stability which can be explained by considering the electronic configuration.

Element	1st Ionisation Energy (kJ mol^{-1})	2nd Ionisation Energy (kJ mol^{-1})
Na	496	4562
Mg	738	1451

> **First and second ionisation energies** Go online
>
> Ionisation energy information provides valuable evidence for the electronic configuration of atoms and the position of elements in the periodic table.
>
> **Q23:** Complete the following paragraph by putting the correct words from the list in place:
>
> Sodium has .... in its outer shell whereas magnesium has .... in its outer shell. The first ionisation energy of magnesium is .... than that of sodium since magnesium has 12 protons in its nucleus and therefore has a higher nuclear charge and a .... attraction for the outer electrons. However, the second ionisation energy of .... is higher than that of .... since the electrons being removed come from a .... p subshell which is .... to the nucleus.
>
> *Word list*: closer, complete, higher, magnesium, one electron, sodium, stronger, two electrons.

2.8 Covalent bonding

A covalent bond is formed when atomic orbitals overlap to form a molecular orbital. When a molecular orbital is formed, it creates a covalent bond. Pure covalent bonding (non-polar) and ionic bonding can be considered to be at opposite ends of a bonding continuum. Polar covalent bonding sits between these two extremes.

Valence shell electron pair repulsion (VSEPR) theory does not provide an accurate description of the actual molecular orbitals in a molecule. However, the shapes of molecules and polyatomic ions predicted are usually quite accurate.

The following figure shows the electrostatic forces in a hydrogen molecule. The positively charged nuclei will repel each other, as will the negatively charged electrons; but these forces are more than balanced by the attraction between the nuclei and electrons.

Figure 2.12: Attractive and repulsive forces in a hydrogen molecule

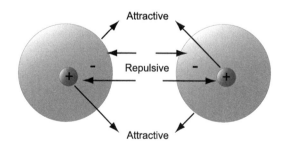

TOPIC 2. ATOMIC ORBITALS, ELECTRONIC CONFIGURATIONS AND THE PERIODIC TABLE

As the two Hydrogen atoms approach each other their atomic orbitals overlap and merge to form a molecular orbital forming a covalent bond between the two atoms.

The following graph relates distance between the atoms and potential energy.

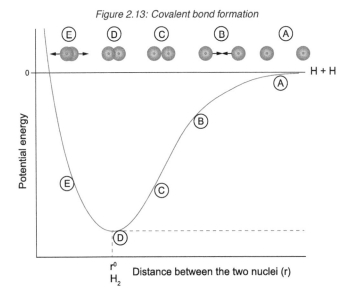

Figure 2.13: Covalent bond formation

At the right hand side of the graph (the start) the two hydrogen atoms are far apart and not interacting (point A on the graph). As the hydrogen atoms approach each other, the nucleus of one atom attracts the neighbouring electron of the other atom and the potential energy drops. At the point when the two atoms become too close and the nuclei repel, the potential energy rises sharply. At point D on the graph a stable situation is reached where the attraction and repulsion are balanced and the lowest potential energy is reached. This is the point where the atomic orbitals have formed a new molecular orbital and the atoms are a bond length apart (r_0). In order to break these atoms apart the equivalent energy released in making the bond would have to be replaced. The quantity of energy required to break a mole of these bonds is known as the bond enthalpy.

Covalent bonding

Q24: At which point is the system most stable?

....

Q25: What does the symbol $r°$ represent?

....

Q26: Would the progress from point A to point C be exothermic or endothermic?

....

> **Q27:** The value of r° on this graph is 74 pm. Use the data book to look up the covalent radius for hydrogen and explain the apparent difference.
>
> ....
>
> **Q28:** Use the information in the last question to predict the bond length in a molecule of hydrogen chloride.

2.9 Dative covalent bonds

When a covalent bond is formed two atomic orbitals join together to form a molecular orbital. Usually both atomic orbitals are half-filled before they join together. However sometimes one of the atoms can provide both the electrons that form the covalent bond and this is called a **dative** covalent bond. This bond is exactly the same as all other covalent bonds differing only in its formation. An example of this would be when an ammonium ion is formed in solution when an ammonia molecule picks up a hydrogen ion. The hydrogen ion (H^+) has no electrons and therefore cannot contribute an electron to the covalent bond. Both the electrons come from the lone pair on the nitrogen atom in the ammonia molecule. The dative bond is often arrowed.

$NH_3(aq) + H^+(aq) \rightarrow NH_4^+(aq)$

In a Lewis electron dot and cross diagram:

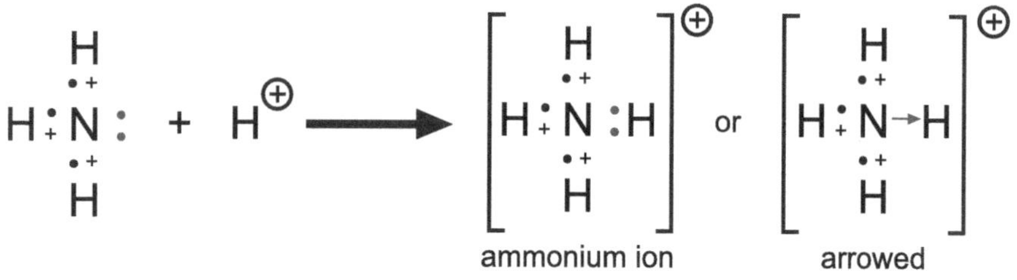

[diagram showing ammonium ion formation with dot-and-cross notation, labelled "ammonium ion" and "arrowed"]

The dative bond is sometimes 'arrowed'.

TOPIC 2. ATOMIC ORBITALS, ELECTRONIC CONFIGURATIONS AND THE PERIODIC TABLE

2.10 Lewis diagrams

Lewis electron dot or dot and cross diagrams (named after American chemist G.N.Lewis) are used to represent bonding and non-bonding pairs in molecules and polyatomic ions.

H ⁑ H   :O⁑⁑O: :N⁑⁑⁑N:

hydrogen   oxygen   nitrogen

Lewis electron dot diagrams

Dots can be used to represent all the electrons or you can use dots to represent only the electrons from one atom and crosses to represent electrons from the other atom. The preceding diagrams show that there is a single covalent bond between the hydrogen atoms, a double bond (2 pairs of electrons) between the oxygen atoms and a triple bond (3 pairs of electrons) between the nitrogen atoms.

Oxygen and nitrogen both have non-bonding electrons which are known as lone pairs (as shown by the outer paired dots). These have an influence on the chemistry of these molecules.

Resonance structures   Go online

Note that resonance structures are not required knowledge and students would not have to draw these from scratch. However, it is entirely possible that resonance structures may be included in the future in a problem solving/skills context.

The Lewis electron dot diagram for ozone O_3 shows there are 6 outer electrons from each oxygen atom giving a total of 18 electrons. There are 2 possible ways to draw the Lewis electron dot diagram for this molecule shown in the following figure. The two different forms are known as resonance structures. The actual structure of the ozone molecule is a hybrid of the two resonance structures also shown in the following diagram.

more easily drawn as:

Another structure with more than one resonance structure is the carbonate ion CO_3^{2-}. There are three different resonance structures for this polyatomic ion.

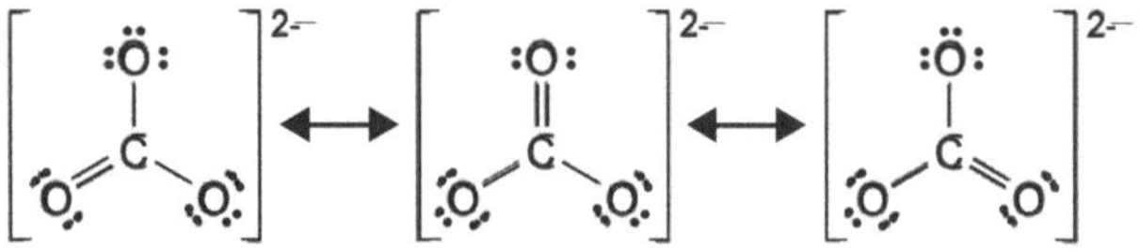

Q29: Which molecule could have this Lewis electron dot diagram?

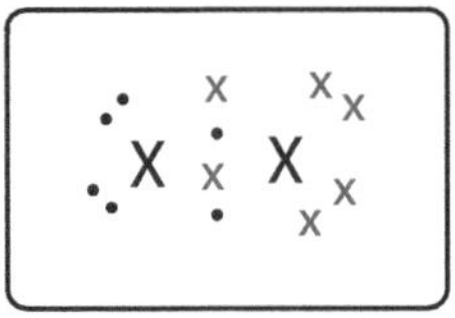

a) Cl_2
b) HCl
c) N_2
d) O_2

....

Q30: Which molecule could be represented by this dot and cross diagram?

a) HCl
b) FCl
c) F_2
d) Cl_2

....

Q31: Draw a Lewis electron dot diagram to show the electrons in: a) methane and b) CO_2.

....

Q32: Carbon monoxide has a structure which contains a double bond and a dative covalent bond from oxygen to carbon. Draw this structure showing the dative bond as an arrow.

TOPIC 2. ATOMIC ORBITALS, ELECTRONIC CONFIGURATIONS AND THE PERIODIC TABLE

2.11 Shapes of molecules and polyatomic ions

Shapes of molecules and polyatomic ions can be predicted by first working out the number of outer electron pairs around the central atom and then dividing them into bonding and non-bonding (lone) pairs. The shape adopted by the molecule or polyatomic ion is one where these electron pairs can be as far apart as possible minimising repulsion between them.

$$\text{electron pairs} = \frac{\text{number of electrons on centre} + \text{number of bonded atoms}}{2}$$

If we apply this to a molecule of ammonia NH_3 which has 5 outer electrons on the central nitrogen atom (electron arrangement 2, 5) + 3 hydrogen atoms bonded:

$$\frac{5+3}{2} = 4 \text{ electron pairs (3 bonded and 1 lone pair)}$$

Total number of electron pairs	Arrangement of electron pairs
2	Linear
3	Trigonal
4	Tetrahedral
5	Trigonal bipyramidal
6	Octahedral

2.12 Examples of molecules with different shapes

Two bonding pairs (BeCl₂)

Beryllium is in group two and therefore has two outer electrons. The two Cl atoms contribute one electron each giving four electrons in two electron pairs. As there are two Cl atoms bonded to the Be, these two electron pairs are bonding electrons and $BeCl_2$ will be a *linear* molecule with bond angles equal to 180°.

Cl - Be - Cl

Three pairs of electrons

Three bonding pairs ($BCl_3(g)$)

Boron is in group 3 and therefore has three electrons in the outer shell. The three Cl atoms provide one electron each giving six electrons in three electron pairs. All three electron pairs are involved in bonding and are therefore bonding electrons. No lone pairs exist on Boron. The BCl_3 will have a *trigonal (trigonal planar)* shape with all four atoms in the same plane.

When this idea is extended to three pairs of electrons as in BCl_3 the molecule is flat with an angle of 120° and is described as trigonal planar.

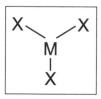

Four bonding pairs (CH_4)

a) Perfect *tetrahedron* with bond angles of 109.5°.

b) Four pairs of electrons with three bonding pairs and one lone pair (NH_3)

Nitrogen is in group five and therefore has five outer electrons. Each H atom provides one electron giving a total of eight electrons (four pairs) around the N atom. Three of the pairs are involved in bonding to the hydrogen atoms leaving one pair as a non-bonding pair (lone pair). Repulsion between a lone pair and a bonding pair is greater than repulsion between bonding pairs. This means the lone pair on the N atom pushes the three nitrogen hydrogen bonds closer together resulting in a slightly smaller bond angle of 107°. This would be described as a **trigonal pyramidal** molecule.

TOPIC 2. ATOMIC ORBITALS, ELECTRONIC CONFIGURATIONS AND THE PERIODIC TABLE

c) Four pairs of electrons with two bonding pairs and two lone pairs (H_2O)

Oxygen is in group six and therefore has six electrons in the outer shell. Each hydrogen atom contributes one electron making a total of eight (four pairs of) electrons around the central O atom. Two of the pairs are involved in bonding with the hydrogen atoms leaving two pairs as non-bonding (lone pairs). The two lone pairs push the two O-H bonds closer together due to greater repulsion between lone pairs giving a bond angle of 104.5°. The shape of this molecule is **angular**.

Five pairs of electrons

Five bonding pairs ($PCl_5(g)$)

Phosphorus has five outer electrons and each of the five chlorine atoms provides one electron giving a total of ten electrons (five pairs). All electron pairs are bonding pairs involved in P-Cl bonds. The shape of this molecule is *trigonal bipyramidal*.

When five pairs of electrons are involved the shape is said to be trigonal bipyramid with angles of 120°, 90° and 180°. An example is PCl_5.

Remember that one or even two of these sites could be occupied by electrons only and the molecule shape would be changed from the trigonal bipyramid.

Six pairs of electrons

Six bonding pairs ($SF_6(g)$)

Sulfur has six outer electrons and each of the six fluorine atoms provides one electron giving a total of 12 electrons (six pairs). All electron pairs are bonding pairs involved in S-F bonds. The shape of this molecule is *octahedral*.

Sulfur hexafluoride

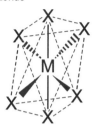

Six pairs of electrons is the highest number we are likely to encounter. For example in SF_6. Bond angles are 90°. Such a structure is described as octahedral. Octahedral geometry is relatively common.

One other common structure worth mentioning is that resulting from an octahedral arrangement which involves two lone pairs.

The iodine tetrachloride negative ion

Six pairs of electrons with two of these as lone pairs results in their "repulsive power" keeping those two the furthest apart. For example in ICl_4^-. Bond angles are 90°. Such a structure is described as *square planar*.

The negative charge in the iodine tetrachloride negative ion adds one electron to the total number. (Similarly, a positive charge on an ion removes one electron from the total.)

A lone pair of non bonding electrons is more repulsive than a bonded pair. The different strength of electron pair repulsion accounts for slight deviations from the expected bond angles in a number of molecules.

109.5° 107° 104.5°

Electron pair repulsions decrease in strength in the order non-bonding pair/non-bonding pair > non-bonding pair/bonding pair > bonding pair/bonding pair.

TOPIC 2. ATOMIC ORBITALS, ELECTRONIC CONFIGURATIONS AND THE PERIODIC TABLE

Summary of shapes of covalent molecules

Go online

Q33: Complete the following table using the items from the list.

Number of Electron Pairs	Arrangement	Angle(s) in degrees	Example
2	linear		
3			BF_3
4	tetrahedral		
5		90, 120, 180	
6			SF_6

Item list: 90, 109.5, 120, 180, $BeCl_2$, CH_4, octahedral, PCl_5, trigonal bipyramidal, trigonal planar.

Q34: Which of these molecules has a non-bonded electron pair on the central atom?

a) PF_3
b) BF_3
c) $BeCl_2$
d) H_2

....

Q35: What is the likely structure of an antimony (V) chloride molecule?

a) Linear
b) Tetrahedral
c) Trigonal bipyramid
d) Octahedral

....

Q36: Which of these would have a bond angle greater than 109.5°?

a) CCl_4
b) NH_3
c) SCl_2
d) BeF_2

....

Q37: Sulfur has six outer shell electrons. Draw a diagram to show the SF_5^- ion structure (the negative ion).

....

Q38: What shape describes the arrangement of the electron pairs you have just drawn?

....

Q39: Suggest a name for the shape of this **molecule** (remember to ignore any lone pairs).

2.13 Summary

Summary

You should now be able to:

- relate spectral evidence to electron movements and ionisation energy;
- describe the four quantum numbers and relate these to atomic orbitals, their shape and relative energies;
- describe the four s, p and d atomic orbitals, their shapes and relative energies;
- relate the ionisation energies of elements to their electronic configuration, and therefore to their position in the periodic table;
- describe the electronic configuration of atoms 1-20 in spectroscopic notation;
- write electron configurations in spectroscopic notation;
- explain that covalent bonding involves the sharing of electrons and can describe this through the use of Lewis electron dot diagrams;
- predict the shape of molecules and polyatomic ions through consideration of bonding pairs and non-bonding pairs and the repulsion between them;
- understand the decreasing strength of the degree of repulsion from lone-pair/lone-pair to non-bonding/bonding pair to bonding pair/bonding pair.

2.14 Resources

- Royal Society of Chemistry (http://www.rsc.org)
- SciComm video (https://youtu.be/K-jNgq16jEY) - 3D models of orbitals.
- Chemical Education Digital Library (http://bit.ly/29Jikva)
- Animated Molecules (http://bit.ly/29XBA76)
- Chemistry Pages http://bit.ly/29Sgu9O

TOPIC 2. ATOMIC ORBITALS, ELECTRONIC CONFIGURATIONS AND THE PERIODIC TABLE

2.15 End of topic test

End of Topic 2 test Go online

The end of topic test for *Atomic orbitals, electronic configurations and the periodic table*

Q40: 'Each atomic orbital can hold a maximum of only two electrons.'
This is a statement of:

a) the Pauli exclusion principle.
b) Hund's rule.
c) the aufbau principle.
d) the Heisenberg uncertainty principle.

....

Q41: Which element has the following spectroscopic notation?

[Ne] $3s^2\ 3p^5$

a) Neon
b) Boron
c) Nitrogen
d) Chlorine

....

Q42: How many quantum numbers are necessary to identify an electron in an atomic orbital?

a) 1
b) 2
c) 3
d) 4

....

Q43: Which of these electron arrangements breaks Hund's rule?

a)
b)
c)
d)

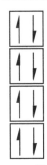
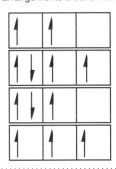

....

Q44: In the emission spectra of hydrogen, how many lines are produced by electron transitions involving only the three lowest energy levels?

a) 1
b) 2
c) 3
d) 4

....

Q45: The first and second ionisation energies of boron are 801 and 2427 kJ mol^{-1} respectively.

This means that for one mole of gaseous boron 3228 kJ of energy:

a) would be needed to remove 2 moles of 2s electrons.
b) would be needed to remove 1 mole of 2p electrons and 1 mole of 2s electrons.
c) would be released when 1 mole of 2p electrons and 1 mole of 2s electrons are removed.
d) would be released when 2 moles of 2s electrons are removed.

....

Q46: The second ionisation energy of magnesium can be represented by:

a) $Mg^+(g) \rightarrow Mg^{2+}(g) + e^-$
b) $Mg(s) \rightarrow Mg^{2+}(g) + 2e^-$
c) $Mg(g) \rightarrow Mg^{2+}(g) + 2e^-$
d) $Mg^{2+}(g) \rightarrow Mg^{3+}(g) + e.$

....

Q47: What is the total number of electrons which may occupy a p sub-shell and remain unpaired?

....

Q48: How many unpaired electrons are there in a fluorine atom?

....

Q49: What is the maximum number of quantum numbers which can be the same for any two electrons in an atom?

....

Q50: What word is used to describe the orbitals within the p, d or f sub-shells which have the same energy?

....

Q51: Which element has atoms with the same spectroscopic notation as a Calcium ion (Ca^{2+})?

....

Q52: Explain why the lines in an emission spectrum become closer and closer together as they converge towards the high energy end of the spectrum for an element.

....

Q53: The first ionisation energies for the p-block elements aluminium to argon follow an upward trend, with the exception of phosphorus.

Explain this in terms of the electronic configurations of phosphorus and sulfur.

....

Q54: Which of these compounds has the greatest degree of ionic character?

a) Beryllium oxide
b) Beryllium sulfide
c) Magnesium oxide
d) Calcium oxide

....

Q55: Forming a dative covalent bond between the phosphorus of PH_3 and the boron in BF_3 involves:

a) phosphorus losing electrons to boron.
b) boron losing electrons to phosphorus.
c) reducing the number of electrons in the boron outer shell.
d) phosphorus donating both electrons of the bond to boron.

....

Q56: Which of these is a non-linear molecule?

a) CO
b) CO_2
c) H_2S
d) $BeCl_2$

....

Q57: What change occurs in the three-dimensional arrangement of bonds around the boron in this reaction?

$BF_3 + F^- \rightarrow BF_4^-$

a) Trigonal planar to pyramidal
b) Trigonal planar to tetrahedral
c) Pyramidal to tetrahedral
d) Pyramidal to square planar

....

Q58: The bond angle in a molecule of ammonia is:

a) 90°
b) 107°
c) 109.5°
d) 120°

....

Q59: Identify the shape of the F_2O molecule.

a) Linear
b) Angular
c) Trigonal planar
d) Trigonal pyramidal
e) Tetrahedral
f) Trigonal bipyramidal

....

Q60: Identify the shape of the nitrogen fluoride molecule.

a) Linear
b) Angular
c) Trigonal planar
d) Trigonal pyramidal
e) Tetrahedral
f) Trigonal bipyramidal

....

Q61: Identify the **two** three-dimensional shapes around the oxygen atom in this reaction:
$H_2O + H^+ \rightarrow H_3O^+$

a) Linear
b) Angular
c) Trigonal planar
d) Trigonal pyramidal
e) Tetrahedral
f) Trigonal bipyramidal

This diagram shows the outer electron arrangements in a polyatomic ion.

The oxygen atoms are labelled A, B, C, D.

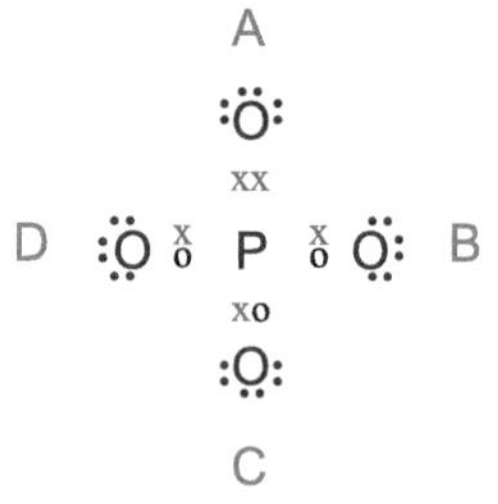

```
            A
           :Ö:
           xx
  D  :Ö: x/o  P  x/o :Ö:  B
           xo
           :Ö:
            C
```

TOPIC 2. ATOMIC ORBITALS, ELECTRONIC CONFIGURATIONS AND THE PERIODIC TABLE

Q62: Which of these describes the shape of the ion?

a) Tetrahedral
b) Trigonal pyramidal
c) Square planar
d) Trigonal biplanar

....

Q63: What charge would this ion carry?

a) One negative
b) Two negative
c) Three negative
d) Five negative

....

Q64: Which of the oxygen atoms is attached by a dative covalent bond?

a) A
b) B
c) C
d) D

....

Q65: The following image shows a dot and cross diagram of how a dative covalent bond is formed in this reaction:

$BF_3 + F^- \rightarrow BF_4^-$

Identify the dative covalent bond.

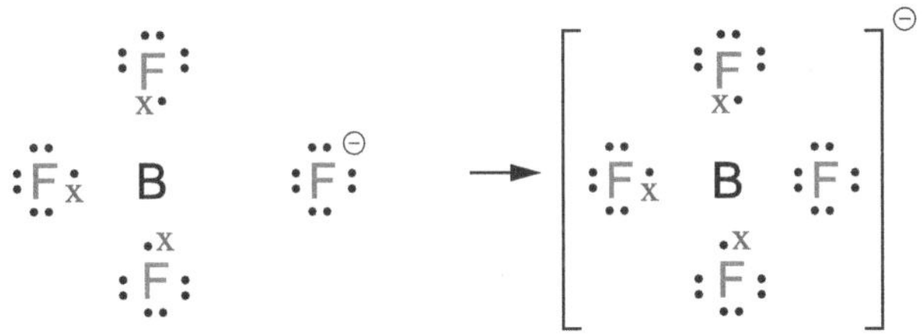

The noble gas xenon can combine with fluorine under certain circumstances. Xenon tetrafluoride is one possible product.

Q66: Which of the following shows a sketch of the xenon tetrafluoride molecule?

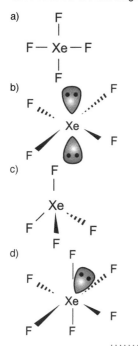

Q67: Name this shape. ....

Unit 1 Topic 3

Transition metals

Contents

- 3.1 Electronic configuration 57
- 3.2 Oxidation states and oxidation numbers 58
    - 3.2.1 Calculating an oxidation state 59
    - 3.2.2 Multiple oxidation states 60
    - 3.2.3 Oxidation and reduction 60
- 3.3 Ligands and transition metal complexes 61
    - 3.3.1 Coordination number and shape of ligands 62
    - 3.3.2 Naming complexes 63
- 3.4 Colour in transition metal complexes 66
- 3.5 UV and visible spectroscopy 69
- 3.6 Catalysis 74
- 3.7 Summary 76
- 3.8 Resources 76
- 3.9 End of topic test 77

Learning objective

By the end of this topic, you should be able to:

- understand and draw electronic configuration diagrams for transition metal atoms and ions;
- understand and explain any anomalies in the electronic configuration model;
- work out the oxidation state of transition metals and the oxidation number of transition metal ions;
- explain that changes in oxidation number show oxidation and reduction reactions;
- understand what allows a substance to be used as a ligand and how their classification and the coordination number are worked out;
- name complex ions according to IUPAC rules;
- explain what causes transition metal complexes to be coloured;
- understand UV and visible absorption spectroscopy of transition metal complexes;
- understand that transition metal complexes can be used in catalysis.

3.1 Electronic configuration

Transition metals are found between groups 2 and 3 on the periodic table and are known as the d block elements. They have many important uses including piping, electrical wiring, coinage, construction and jewellery. Many have important biological uses and many are used as industrial catalysts.

The d block transition metals are metals with an incomplete d subshell in at least one of their ions. This gives transition metals their distinctive properties and we will be concentrating on the first row of transition metals from scandium to zinc.

As we go across the row from scandium to zinc the transition metals follow the aufbau principle, adding electrons to the subshells one at a time in order of their increasing energy, starting with the lowest. This must fit in with the electron arrangement given in the SQA data booklet.

Scandium has the electronic configuration $1s^2 2s^2 2p^6 3s^2 3p^6 3d^1 4s^2$

The 4s orbital has been filled before the 3d orbital due to being lower in energy.

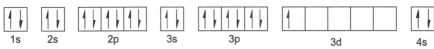

Electronic configuration of scandium written in orbital box notation

This can be shortened to [Ar] $3d^1$ $4s^2$ where [Ar] represents the s and p orbitals of the Argon core.

Copper and chromium appear not to follow the aufbau principle (orbitals are filled in order of increasing energy).

Chromium [Ar] $3d^5$ $4s^1$

Copper [Ar] $3d^{10}$ $4s^1$

Half-filled or fully filled d orbitals have a special stability. ***However, whenever transition metals form ions, electrons are lost first from the outermost subshell the 4s.***

Electronic configuration of Co^{2+} is therefore [Ar] $3d^7$

Orbital box notation

Go online

Q1: Using the orbital box, practise working out the orbital box notations for the transition metals scandium, titanium, vanadium, chromium, manganese, iron, cobalt, nickel, copper and zinc.

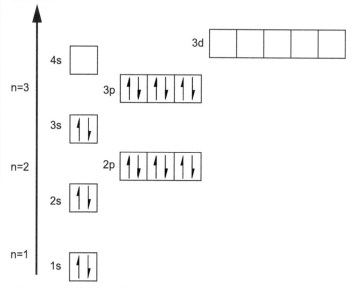

Electronic configuration

Q2: Explain why scandium and zinc are often considered not to be transition metals.

....

Q3: Consider the electronic configurations of the Fe^{2+} and Fe^{3+} ions in terms of orbital box notation. Explain why Fe(III) compounds are more stable than Fe(II) compounds.

3.2 Oxidation states and oxidation numbers

The oxidation state is similar to the valency that an element has when it is part of a compound. Iron(II) chloride would normally be stated as having iron with a valency of 2, but it is actually more accurate to say that the iron is in an oxidation state (II) or has oxidation number +2.

Rules need to be followed when assigning an oxidation number to an element.

TOPIC 3. TRANSITION METALS

Rule No.	Rule
1.	Oxidation number of an uncombined element is 0.
2.	For ions containing single atoms (monatomic) the oxidation number is the same as the charge on the ion. Example Na^+ and Cl^- the oxidation number would be +1 and -1 respectively.
3.	In most compounds oxygen has oxidation number -2.
4.	In most compounds hydrogen has the oxidation number +1. The exception is in metallic hydrides where it is -1.
5.	The sum of all the oxidation numbers of all the atoms in a molecule or neutral compound must add up to 0.
6.	The sum of all the oxidation numbers of all the atoms in a polyatomic ion must add up to the charge on the ion.

3.2.1 Calculating an oxidation state

Oxidation number of Mn in MnO_4^-

We must apply rule 7 here where all the oxidation numbers of the atoms must add up to -1 (charge on the ion). Each oxygen atom has an oxidation number of -2 (rule 3) so the sum of the oxidation numbers on oxygen is $4 \times -2 = -8$. Therefore the oxidation number of Mn must be 7 (-8 + 7 = -1).

Calculating an oxidation state

Find the oxidation number for the transition metal in the following examples.

Q4: VO_2^+

....

Q5: CrO_4^{2-}

....

Q6: $VOCl_2$

....

Q7: Cr_2O_3

....

Q8: $K_2[Cr_2O_7]$

....

Q9: $Na_4[NiCl_6]$

....

Q10: $[FeO_4^{2-}]$

....

© HERIOT-WATT UNIVERSITY

> **Q11:** $K_2[MnO_4]$
>
> ....
>
> **Q12:** $K_3[CoF_6]$

3.2.2 Multiple oxidation states

Transition metals may have more than one oxidation state in their compounds. Iron for example has the familiar oxidation states of (II) and (III). Copper is predominately in oxidation state (II) but can have an oxidation number of +1 in Cu_2O.

Transition metal compounds can exhibit different colours depending on the oxidation state of the metal. For example iron(II) compounds are often pale green and iron(III) compounds are yellow-orange. Iron(II) compounds are less stable than iron(III) since the iron(II) becomes slowly oxidised to iron(III).

The relative stabilities of the different oxidation states are determined by several factors including:

- the electronic structure (which influences ionisation energies and ionic radius);
- the type of bonding involved;
- the stereochemistry.

Less common oxidation states are shown in brackets.

				7+					
			6+	(6+)	(6+)				
		5+							
	4+	4+		4+					
3+	(3+)	(3+)	3+	3+	3+	3+	(3+)		
	(2+)	(2+)	(2+)	2+	2+	2+	2+	2+	2+
								1+	
Sc	Ti	V	Cr	Mn	Fe	Co	Ni	Cu	Zn

Common oxidation states of the first transition metal series.

3.2.3 Oxidation and reduction

OIL - Oxidation is a loss of electrons

This can also be also shown as an increase in the oxidation number of the transition metal.

RIG - Reduction is a gain of electrons

This can also be shown as a decrease in the oxidation number of the transition metal.

Determine if the conversion from VO^{2+} to VO_2^+ is oxidation or reduction.

TOPIC 3. TRANSITION METALS

VO^{2+}	VO_2^+
Overall charge of ion = +2	Overall charge of ion = +1
Oxidation number of O = -2	Oxidation number of O = 2 × (-2) = -4
Oxidation number of V = +2 - (-2) = +4	Oxidation number of V = +1 - (-4) = +5

The oxidation number of vanadium has increased from +4 in VO^{2+} to +5 in VO_2^+ showing it has been oxidised.

Compounds containing metals in high oxidation states tend to be oxidising agents whereas those containing metals in low oxidation states tend to be reducing agents.

3.3 Ligands and transition metal complexes

Ligands are electron donors which are usually negative ions or molecules that have one or more non-bonding (lone) pairs of electrons. When these ligands surround a central transition metal ion they form a transition metal complex often called a coordination compound.

Chloride ion	Cyanide ion	Ammonia molecule	Water molecule
:Cl:⁻	:C≡N:⁻	H–N(H)(H) with lone pair	H–O(H) with lone pairs

Monodentate ligands

These ligands are known as monodentate which means they donate one pair of electrons to the central transition metal ion i.e. form a dative bond. A bidentate ligand donates two pairs of electrons to the central transition metal ion and examples include the oxalate ion and 1, 2-diaminoethane (ethylene diamine abbreviated to 'en').

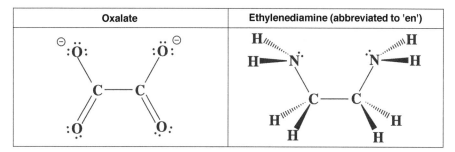

Bidentate ligands

© HERIOT-WATT UNIVERSITY

EDTA (ethylenediaminetetraacetate) is a hexadenate ligand as it has 6 non-bonding pairs of electrons which bond to the central transition metal ion. EDTA reacts with metal ions such as Ni^{2+} in a 1:1 ratio.

EDTA^{4-}	Nickel EDTA complex
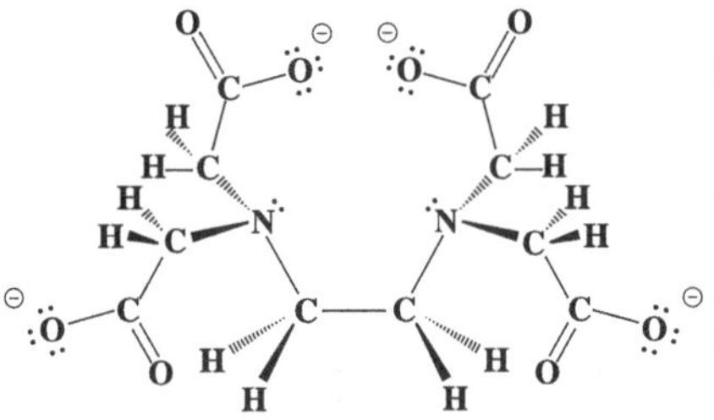 (structure of EDTA^{4-})	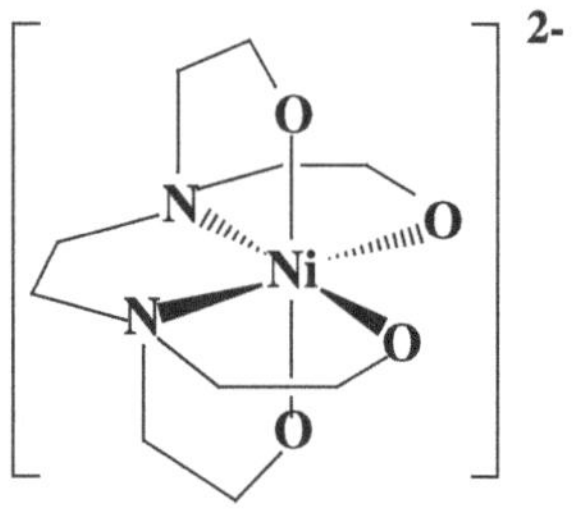 (structure of Nickel EDTA complex, charge 2-)

3.3.1 Coordination number and shape of ligands

The **coordination number** of the central transition metal ion is the number of bonds from it to the ligands. The coordination number will determine the shape of the complex ion.

Coordination number	Shape	
2	Linear	X—M—X
4	Square planar	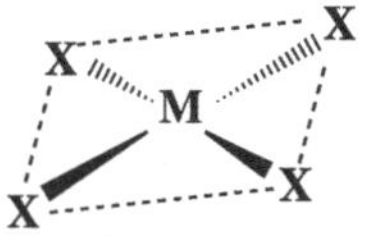 (square planar diagram)
4	Tetrahedral	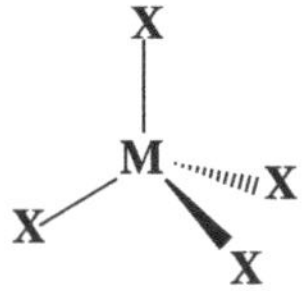 (tetrahedral diagram)
6	Octahedral	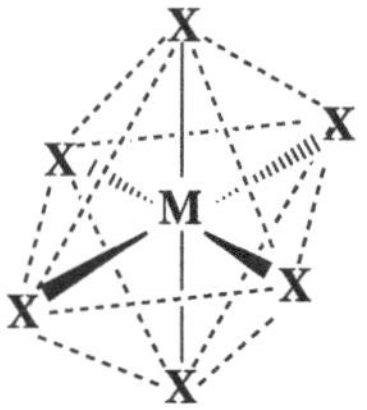 (octahedral diagram)

Coordination number and shape

3.3.2 Naming complexes

Transition metal complexes are named and written according to IUPAC rules.

Rule No.	Rule
1	The symbol of the metal is written first, followed by the symbols of the ligands in alphabetical order according to which atom of the ligand binds. For water as a ligand, OH_2 is used, rather than H_2O, since it is the oxygen atom that binds to the metal. Similarly, for oxalate, $O_2C_2O_2$ is used in the formula, rather than C_2O_4.
2	Formula of the complex ion is enclosed within square brackets $[FeCl_2(OH_2)_4]^+$ with the charge outside the square brackets.
3	Ligands are named in alphabetical order followed by the name of the metal and its oxidation state. If there is more than one of a ligand it is preceded by the prefix for the number di, tri, tetra etc.
4	If the ligand is a negative ion ending in -ide then in the complex name the ligand name changes to end in 'o'. Chloride become chlorido and cyanide becomes cyanido.
5	If the ligand is ammonia NH_3 it is named as ammine. Water as a ligand is named aqua.
6	If the complex is a negative ion overall the name of the complex ends in -ate. Cobaltate would be for a negative ion containing cobalt. However for copper cuprate is used and ferrate for iron.
7	If the complex is a salt the name of the positive ion precedes the name of the negative ion.

Some common ligands and their names are as follows.

- Ammonia, NH_3 (ammine)
- Bromide, Br^- (bromido)
- Chloride, Cl^- (chlorido)
- Cyanide, CN^- (cyanido)
- Fluoride, F^- (fluorido)
- Hydroxide, OH^- (hydroxido)
- Iodide, I^- (iodido)
- Oxalate, $C_2O_4^{2-}$ (oxalato)
- Water, H_2O (aqua)

Examples of naming complex ions

$[Cu(OH_2)_4]^{2+}$ is named tetraaquacopper (II)

$[Co(NH_3)_6]^{2+}$ is named hexaamminecobalt (II).

$[Fe(CN)_6]^{4-}$ is named hexacyanidoferrate (II).

If we have $K_3[Fe(CN)_6]$ this would be called potassium hexacyanidoferrate (III).

$(K^+)_3$ each K has a 1+ charge so three would contribute a 3+ charge. This means the negative ion from the complex will have an overall 3- charge. Each cyanide ion contributes a -1 charge so six of them would contribute a -6 charge. This means the oxidation state of iron would be 3+.

$(CN)_6 = -6$

Overall charge on negative ion = -3

Oxidation state of Fe = (-3 + -6) = +3

Naming transition metal complexes   Go online

Part 1

For each of the following complexes, write the correct name. Be very careful to spell each part of the name accurately (no capital letters) and don't put in spaces unless they are needed.

Q13: $[Co(OH_2)_6]Cl_2$

....

Q14: $Na[CrF_4]$

....

Q15: $K_4[Fe(CN)_6]$

....

Q16: $K_3[Fe(O_2C_2O_2)_3]$

Part 2

For each of the following compounds what is the coordination number of the transition metal ion?

Q17: $Na[CrF_4]$

....

Q18: $K_3[Fe(O_2C_2O_2)_3]$

....

Q19: $K_4[Fe(CN)_6]$

....

Q20: Predict the shape of the complex ion in the previous question.

TOPIC 3. TRANSITION METALS

Part 3

What is the correct structural formula for each of the following compounds?

Q21: Sodium tetrachloridoplatinate (II)

a) Na[PtCl$_4$]
b) Na$_2$[PtCl$_4$]
c) Na[Pt$_2$Cl$_4$]
d) Na$_4$[PtCl$_4$]

....

Q22: Diaquadicyanidocopper (II)

a) (H$_2$O)$_2$(CN)$_2$Cu
b) [Cu(CN$_2$)(H$_2$O$_2$)]
c) Cu(CN)(OH$_2$)$_2$
d) [Cu(CN)$_2$(OH$_2$)$_2$]

....

Q23: Pentaaquachloridochromium (III) chloride

a) [CrCl(OH$_2$)$_5$]Cl$_2$
b) [Cr(OH$_2$)$_5$]Cl$_3$
c) [CrCl(OH$_2$)$_5$]Cl$_3$
d) [CrCl$_5$(OH$_2$)$_5$]Cl

....

Q24: Tetraamminedichloridocobalt (III) chloride

a) [CoCl$_2$(NH$_3$)$_4$]Cl$_3$
b) [CoCl$_2$(NH$_3$)$_4$]Cl$_2$
c) [CoCl$_2$(NH$_3$)$_4$]Cl
d) [Co(NH$_3$)$_4$]Cl$_3$

3.4 Colour in transition metal complexes

Several transition metal complexes are coloured including solutions of copper(II) compounds which are blue and solutions of nickel(II) complexes are green. To explain how these colours arise we need to look at the identity and oxidation state of the transition metal and the ligands attached in the complex.

White light is a complete spectrum ranging from 400 to 700nm known as the visible region of the electromagnetic spectrum. White light consists of all the colours of the rainbow. A complex appears coloured when some of this spectrum is absorbed and colourless when none is absorbed. If all the colours are absorbed the complex will appear black.

Transition metal complexes are able to absorb light due to the five degenerate d orbitals splitting in terms of energy. In a free transition metal ion (one without ligands) the five d orbitals in the 3d subshell are degenerate (equal in energy). On the formation of a complex for example $[Ni(H_2O)_6]^{2+}$ six water ligands surround the central nickel ion forming an octahedral shaped complex. The ligands approach the nickel ion along the x,y and z axes. The electrons in d orbitals that lie along these axes (dz^2 and dx^2-y^2) will be repelled by electrons in the water ligand molecules. These orbitals now have higher energy than the three d orbitals that lie between the axes (dxy, dyz and dxz) and therefore the five d orbitals are no longer degenerate. This is called splitting of d orbitals and is different in octahedral complexes compared to tetrahedral and other shapes of complexes.

The energy difference between the different subsets of d orbitals depends on the ligand and its position in the **spectrochemical series** (series of order of ligand's ability to split the d orbitals).

$$CN^- > NH_3 > H_2O > OH^- > F^- > Cl^- > Br^- > I^-$$

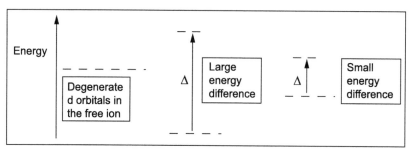

The difference in energy between the two subsets of d orbitals is known as the crystal field strength. This is given the symbol delta Δ.

Colour of transition metal compounds

Go online

Part 1

White light consists of all the wavelengths of light in the visible spectrum combined. When compounds absorb radiation from the visible spectrum, the colour corresponding to this wavelength is removed from the white light and the colour that remains is the complementary colour. The colour wheel shows complementary colours of light, i.e. if green light is absorbed then purple light is transmitted.

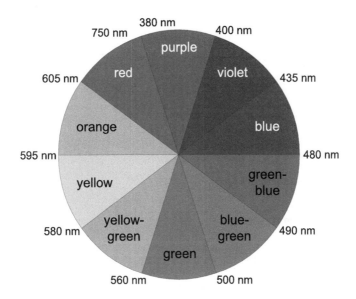

Figure 3.1: Colour wheel

Q25: Complete the following passage using words from the list.

If violet is absorbed, ... is transmitted.

If .... is absorbed, green is transmitted.

If orange is absorbed, ... is transmitted.

When all colours of light are present .... light is produced.

Word list: blue, blue-green, green, green-blue, orange, purple, red, yellow, yellow-green, violet, white.

Part 2

Figure 3.2: Absorption of colour by a complex

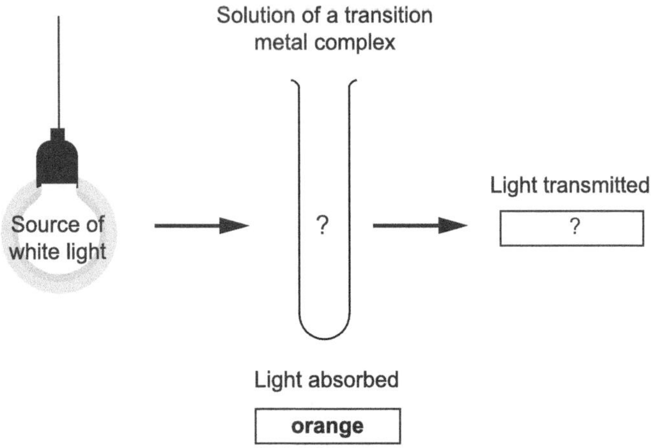

Solution of a transition metal complex

Source of white light → ? → Light transmitted: ?

Light absorbed

| orange |

In the image, white light is passed through a solution of a transition metal complex. Some visible light is absorbed.

Q26: What colour of light is transmitted?

....

Q27: What colour is the solution?

....

Q28: If violet light had been absorbed, what colour would the solution have appeared?

When one particular colour of light is absorbed, the colour remaining is the **complementary colour**. In other words the transmitted light is the complementary colour of the absorbed light.

Key point

Compounds are coloured because they absorb radiation from the visible part of the spectrum. The colour of a compound is that of the light which is not absorbed.

3.5 UV and visible spectroscopy

Transition metal complexes absorb light due to the split in d orbitals. Electrons in the lower d orbitals can absorb energy and move to the higher energy d orbitals. If this energy absorbed in this d-d transition is in the visible region of the electromagnetic spectrum the colour of the transition metal complex will be the complementary colour of the colour absorbed.

The effects of d-d transitions can be studied using spectroscopy. If the absorbed energy is in the visible part of the electromagnetic spectrum (400-700 nm) the complex will be coloured and visible spectroscopy would be used. If the absorbed energy is in the UV part of the electromagnetic spectrum (200-400 nm) the transition metal complex will be colourless and UV spectroscopy will be used.

If the ligands surrounding the transition metal ion are strong field ligands (those that cause the greatest splitting of the d orbitals) d-d transitions are more likely to occur in the UV region of the electromagnetic spectrum. If the ligands are weak field ligands (those that split the d orbitals least) the energy absorbed is more likely to occur in the visible region of the electromagnetic spectrum. These complexes will be coloured.

A colorimeter fitted with coloured filters corresponding to certain wavelengths in the visible region can be used to measure the absorbance of coloured solutions. A filter of the complementary colour should be used.

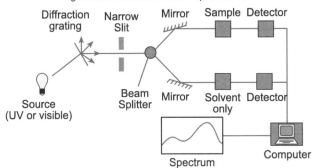

Figure 3.3: Ultraviolet / visible spectrometer

Samples are used in solution and are placed in a cell. Another identical cell containing the pure solvent is also placed in the machine. Radiation across the whole range is scanned continuously through both the sample solution and the pure solvent. The spectrometer compares the two beams. The difference is the light absorbed by the compound in the sample. This data is produced as a graph of wavelength against absorbance. An example is shown in the following figure.

Figure 3.4: Ti^{3+}(aq) visible spectrum

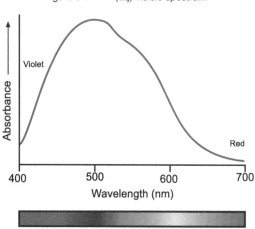

A UV spectrometer passes different wavelengths of UV light from 200 to 400nm through a sample and the quantity of UV light absorbed at different wavelengths is recorded. The intensity of the light absorbed at a given wavelength (especially the wavelength of maximum absorbance) is proportional to concentration, therefore UV/visible spectrum can also be used for quantitative analysis (colorimetry).

Explanation of colour in transition metal compounds   Go online

In this activity, three chromium(III) complexes will be considered. All are octahedral complexes which differ only in the nature of the ligands surrounding the central chromium(III) ion. The chromium(III) ion has a d^3 configuration. In an octahedral complex the d orbitals will be split and absorption of energy in the visible region can promote an electron from the lower to the higher level. When electrons are promoted to higher energy levels, the colour seen is entirely due to this absorption and the complementary colour being transmitted.

Note that no photon emission is involved as the electrons do not fall back down to lower energy levels. This is a very common misconception by students who often confuse the two concepts. Generally emission will only occur during a flame test (or after a high energy electrical spark). If emission did occur, then the colour seen would be the same wavelength as the colour absorbed, i.e. not the complementary colour.

Figure 3.5: Splitting of d orbitals in an octahedral complex

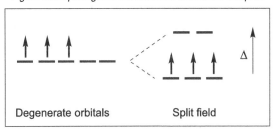

Figure 3.6: Absorption of a photon of light by an octahedral complex

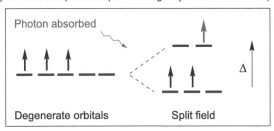

The hexachloridochromate (III) ion formula: $[CrCl_6]^{3-}$

Figure 3.7: Visible spectrum of hexachloridochromate (III) ion: note non linear scale on x axis

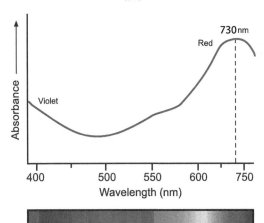

Q29: Use the wavelength of the most intense absorption in the visible spectrum (see the preceding figure) to calculate in kJ mol^{-1} the crystal field splitting (Δ) caused by the chloride ion (give your answer to one decimal place).

....

Q30: What colour would you predict for a solution containing $[CrCl_6]^{3-}$ ions?

a) Red
b) Blue-green
c) Violet
d) Yellow-green

The hexaaquachromium(III) ion formula: $[Cr(OH_2)_6]^{3+}$

Figure 3.8: Visible spectrum of hexaaquachromium(III) ion: note non linear scale on x axis

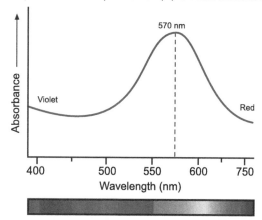

Q31: Use the wavelength of the most intense absorption in the visible spectrum (see the preceding figure) to calculate in kJ mol^{-1} the crystal field splitting (Δ) caused by the water ligand (give your answer to one decimal place).

....

Q32: What colour would you predict for a solution containing $[Cr(OH_2)_6]^{3+}$ ions?

a) Red
b) Yellow
c) Violet
d) Green

The hexaamminechromium(III) ion formula: $[Cr(NH_3)_6]^{3+}$

Figure 3.9: Visible spectrum of hexaamminechromium(III) ion: note non linear scale on x axis

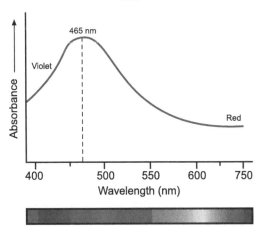

Q33: Use the wavelength of the most intense absorption in the visible spectrum (see the preceding figure) to calculate in kJ mol^{-1} the crystal field splitting (Δ) caused by the ammonia ligand (give your answer to one decimal place).

....

Q34: What colour would you predict for a solution containing $[Cr(NH_3)_6]^{3+}$ ions?

a) Red
b) Yellow
c) Violet
d) Green

....

Q35: The ligands can be placed in order of the crystal field splitting (Δ) with the ligand of lowest energy first. Which of the following shows the correct order?

a) $NH_3 < H_2O < Cl^-$
b) $NH_3 < Cl^- < H_2O$
c) $Cl^- < H_2O < NH_3$
d) $Cl^- < NH_3 < H_2O$

> **Key point**
>
> Different ligands produce different crystal field splittings and so complexes of the same metal ion with different ligands will have different colours.

3.6 Catalysis

Transition metals and their compounds are used as catalysts.

Transition Metal	Process
Iron	Haber Process production of ammonia
Platinum	Ostwald Process production of nitric acid
Platinum/Palladium/Rhodium	Catalytic converters
Nickel	Hardening of oil to make margarine
Vanadium	Contact Process production of sulfuric acid

These are examples of heterogeneous catalysts as they are in a different physical state to the reactants. Transition metals such as iron, copper, manganese, cobalt, nickel and chromium are essential for the effective catalytic activity of certain enzymes showing their importance in biological reactions.

Transition metals are thought to be able to act as catalysts due to atoms on the surface of the active sites forming weak bonds with the reactant molecules using partially filled or empty d orbitals forming intermediate complexes. This weakens the covalent bonds within the reactant molecule and since they are now held in a favourable position they are more likely to be attacked by molecules of the other reactant. This provides an alternative pathway with a lower activation energy increasing the rate of reaction.

Hydrogenation of ethene using a nickel catalyst

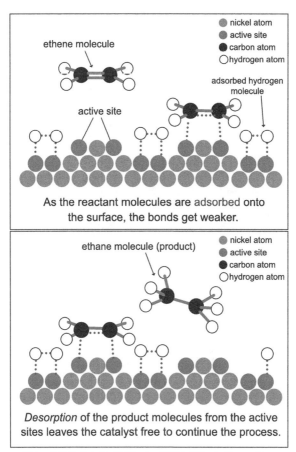

As the reactant molecules are adsorbed onto the surface, the bonds get weaker.

Desorption of the product molecules from the active sites leaves the catalyst free to continue the process.

Note: Adsorption is where something sticks to the surface.

Transition metals are also thought to be able to act as catalysts due to having variable oxidation states. This also allows the transition metal to provide an alternative pathway with a lower activation energy.

Homogenous catalysts (those in the same physical state from the reactants) are used in the reaction of a solution of Rochelle salt (potassium sodium tartrate) and hydrogen peroxide. The catalyst is cobalt(II) chloride solution.

The cobalt(II) chloride solution is pink at the start but changes to green as Co^{3+} ions form. Oxygen gas is vigorously given off at this point. At the end of the reaction Co^{2+} ions are regenerated and the pink colour returns.

3.7 Summary

> **Summary**
>
> You should now be able to state that:
>
> - atoms and ions of the d block transition metals have an incomplete d subshell of electrons;
> - transition metals exhibit variable oxidation states and their chemistry frequently involves redox reactions;
> - transition metals form complexes (coordination compounds) which are named according to IUPAC rules;
> - the properties of these complexes, such as colour, can be explained by the presence of unfilled and partly filled d orbitals;
> - the effects of d → d electronic transitions can be studied using ultraviolet and visible absorption spectroscopy, which is an important analytical tool;
> - transition metals and their compounds are important as catalysts in many reactions, again due to the presence of a partially filled d subshell.

3.8 Resources

- Royal Society of Chemistry (https://rsc.li/2Md570t and https://rsc.li/2TqxNE3)
- Chemguide (http://www.chemguide.co.uk)

3.9 End of topic test

End of Topic 3 test   Go online

Q36: Which of the following electron configurations could represent a transition metal?

a) $1s^2\ 2s^2\ 2p^6\ 3s^2\ 3p^6\ 3d^3\ 4s^2$
b) $1s^2\ 2s^2\ 2p^6\ 3s^2\ 3p^5$
c) $1s^2\ 2s^2\ 2p^6\ 3s^2\ 3p^6\ 3d^{10}\ 4s^2\ 4p^3$
d) $1s^2\ 2s^2\ 2p^6\ 3s^2\ 3p^6\ 4s^2$

....

Q37: Part of the electron configuration of iron can be shown thus:

         3d   4s

Using the same notation, which of the following shows the correct configuration for a chromium atom?

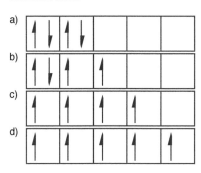   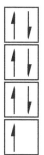

....

Q38: A green hydrated ion has three unpaired electrons. Which of these ions could it be?

a) Fe^{2+}
b) V^{3+}
c) Ni^{2+}
d) Cr^{3+}

....

Q39: What is the oxidation number of nickel in the complex, $Mg_2[NiCl_6]$?

a) +1
b) +2
c) +4
d) +6

Q40: Potassium manganate(VII) (potassium permanganate) is purple in colour. In which region of the visible spectrum does it mainly absorb?

a) Green
b) Blue
c) Yellow
d) Purple

Q41: $VO^{2+} \rightarrow V^{3+}$

This change involves:

a) an oxidation with loss of one electron.
b) a reduction with gain of one electron.
c) an oxidation with loss of three electrons.
d) a reduction with gain of three electrons.

Q42: Which of the following is **true** about ultraviolet spectroscopy?

a) The wavelength range is approximately 400-700 nm.
b) The concentration of the absorbing species can be calculated from the intensity of the absorption.
c) Compounds which absorb only in the ultraviolet are coloured.
d) A UV spectrum is an emission spectrum.

The following three questions refer to a complex which has the formula $[Cr(NH_3)_6]Cl_3$.

Q43: Select **two** terms that can be applied to the complex ion.

a) Cation
b) Tetrahedral
c) Anion
d) Octahedral
e) Hexadentate
f) Monodentate

Q44: Which term can be applied to the ligand?

a) Cation
b) Tetrahedral
c) Anion
d) Octahedral
e) Hexadentate
f) Monodentate

TOPIC 3. TRANSITION METALS

Q45: What is the correct name for this complex?

The $[Co(NH_3)_6]^{3+}$ ion is yellow and the $[CoF_6]^{3-}$ ion is blue.

Q46: What is the oxidation state of cobalt in both complex ions?

....

Q47: What is the name of the ligand that causes the stronger crystal field splitting?

A solution containing hydrogen peroxide and potassium sodium tartrate was heated. No gas was produced. When pink cobalt(II) chloride was added, the solution turned green and bubbles were produced rapidly. As the bubbling subsided, the green colour turned back to pink.

Q48: What evidence is there to suggest that cobalt(II) chloride acts as a catalyst ?

....

Q49: Which of the following statements explains best how this catalyst works?

a) The catalyst provides a surface on which the reaction takes place.
b) Cobalt forms complexes with different colours.
c) Cobalt exhibits various oxidation states of differing stability.
d) The catalyst provides extra energy.

....

Q50: The $[Co(NH_3)_6]^{3+}$ ion is yellow and the $[CoF_6]^{3-}$ ion is blue.

Explain, possibly with the aid of diagrams, why the ions are different colours.

Unit 1 Topic 4

Inorganic Chemistry test

Inorganic Chemistry test

Go online

Q1: Compared to visible radiation (visible light), infra-red radiation has a:

a) lower frequency.
b) higher velocity.
c) lower velocity.
d) higher frequency.

....

Q2: In an emission spectrum the frequency of each line corresponds to:

a) the energy change when an electron moves to a higher energy level.
b) the energy change when an electron moves to a lower energy level.
c) the kinetic energy possessed by an electron in an atom.
d) an energy level within an atom.

....

Q3: The emission spectrum of an element is seen as a series of bright coloured lines against a dark background. The brightest line in the emission spectrum of sodium is seen at 589 nm.

What causes a line in an emission spectrum?

....

Q4: Calculate the frequency of the emission line at 589 nm.

....

Q5: What is the electronic configuration of a vanadium atom?

a) $1s^2 2s^2 2p^6 3s^2 3p^6 3d^3 4s^2$
b) $1s^2 2s^2 2p^6 3s^2 3p^6 3d^4 4s^1$
c) $1s^2 2s^2 2p^6 3s^2 3p^6 3d^5$
d) $1s^2 2s^2 2p^6 3s^2 3p^6 4s^2 4p^3$

....

Q6: When electrons occupy degenerate orbitals, they do so in such a way as to maximise the number of parallel spins.

What is this statement known as?

a) The Pauli exclusion principle
b) The Aufbau principle
c) Hund's rule
d) Heisenberg's uncertainty principle

The 3d and 4s electrons for an iron atom can be represented as follows:

3d    4s

Q7: Complete a corresponding diagram an Fe^{3+} ion.

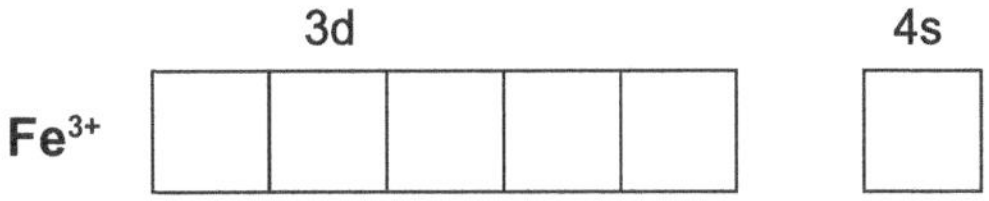

	3d				4s
Fe^{3+}	☐	☐	☐	☐	☐

....

Q8: Suggest why an Fe^{3+} ion is more stable than an Fe^{2+} ion.

....

Q9: What is the shape of an NH_3 molecule?

a) Trigonal pyramidal
b) Trigonal planar
c) Square planar
d) Tetrahedral

....

Q10: An NH_4^+ ion contains four bonding pairs of electrons. What is the shape of the ion?

a) Tetrahedral
b) Trigonal planar
c) Trigonal bipyramidal
d) Linear

Calculate the number of bonding and non-bonding pairs of electrons around the central atom in Cl_2O and hence work out the shape of the molecule.

Q11: How many bonding pairs are there?

....

Q12: How many non-bonding pairs are there?

....

Q13: The shape of the molecule is:

a) angular
b) linear
c) octahedral
d) trigonal pyramidal
e) square planar
f) tetrahedral
g) trigonal planar
h) trigonal bipyramidal

Calculate the number of bonding and non-bonding pairs of electrons around the central atom in PCl_3 and hence work out the shape of the molecule.

Q14: How many bonding pairs are there?

....

Q15: How many non-bonding pairs are there?

....

Q16: The shape of the molecule is:

a) angular
b) linear
c) octahedral
d) trigonal pyramidal
e) square planar
f) tetrahedral
g) trigonal
h) trigonal planar
i) trigonal bipyramidal

Calculate the number of bonding and non-bonding pairs of electrons around the central atom in SiF_4 and hence work out the shape of the molecule.

Q17: How many bonding pairs are there?

....

Q18: How many non-bonding pairs are there?

....

Q19: The shape of the molecule is:

a) angular
b) linear
c) octahedral
d) trigonal pyramidal
e) square planar
f) tetrahedral
g) trigonal
h) trigonal planar
i) trigonal bipyramidal

....

TOPIC 4. INORGANIC CHEMISTRY TEST

Q20: What is the formula for an hexaamminetitanium(III) ion?

a) $[Ti(CH_3NH_2)_6]^{3-}$
b) $[Ti(NH_3)_6]^{3+}$
c) $[Ti(CH_3NH_2)_6]^{3+}$
d) $[Ti(NH_3)_6]^{3-}$

....

Q21: Name the $[CuCl_4]^{2-}$ ion.

....

Q22: What is the oxidation number of iron in the complex ion $[Fe(CN)_6]^{3-}$?

a) -3
b) +2
c) +3
d) +6

Physical Chemistry

1	**Chemical equilibrium**	**89**
1.1	Introduction	91
1.2	Equilibrium expressions and factors affecting equilibrium	94
1.3	Phase equilibria	98
1.4	Acid/base equilibria	104
1.5	Indicators and buffers	115
1.6	Summary	142
1.7	Resources	142
1.8	End of topic test	142
2	**Reaction feasibility**	**153**
2.1	Standard enthalpy of formation	155
2.2	Entropy	155
2.3	Second and third laws of thermodynamics	157
2.4	Free energy	159
2.5	Ellingham diagrams	162
2.6	Summary	168
2.7	Resources	168
2.8	End of topic test	169
3	**Kinetics**	**173**
3.1	Determination of order of reaction	174
3.2	Calculation of rate constants	175
3.3	Reaction mechanisms	177
3.4	Summary	181
3.5	Resources	181
3.6	End of topic test	182
4	**Physical chemistry test**	**187**

Unit 2 Topic 1

Chemical equilibrium

Contents

- 1.1 Introduction ... 91
- 1.2 Equilibrium expressions and factors affecting equilibrium ... 94
- 1.3 Phase equilibria ... 98
    - 1.3.1 Chromatography ... 100
- 1.4 Acid/base equilibria ... 104
    - 1.4.1 Ionisation of water ... 104
    - 1.4.2 pH scale ... 105
    - 1.4.3 Strong/weak acids and bases ... 109
    - 1.4.4 Salts ... 111
- 1.5 Indicators and buffers ... 115
    - 1.5.1 pH titrations ... 115
    - 1.5.2 Buffer solutions ... 122
    - 1.5.3 Calculating pH and buffer composition ... 128
- 1.6 Summary ... 142
- 1.7 Resources ... 142
- 1.8 End of topic test ... 142

Prerequisites

Before you begin this topic, you should be able to:

- state that the forward and backward reactions in dynamic equilibrium have equal rates and that concentrations of products and reactants will remain constant at this time;
- describe how temperature, concentration and pressure affect the position of equilibrium.

Learning objective

By the end of this topic, you should be able to:

- describe the equilibrium chemistry of acids and bases;
- write equilibrium expressions;
- use the terms: pH, K_w, K_a and pK_a;
- understand the chemistry of buffer solutions;
- calculate the pH of buffer solutions.

1.1 Introduction

A chemical reaction is in dynamic equilibrium when the rate of the forward and backward reaction is equal. At this point the concentrations of the reactants and products are constant, but not necessarily equal. From Higher Chemistry you should be aware of the factors that can alter the position of equilibrium including, concentration of reactants or products, pressure and temperature. You should also be aware that a catalyst speeds up the rate at which equilibrium is reached, but does not alter the position of equilibrium.

The nature of chemical equilibrium

A chemical system in equilibrium shows no changes in macroscopic properties, such as overall pressure, total volume and concentration of reactants and products. It appears to be in a completely unchanging state as far as an outside observer is concerned.

Consider a bottle of soda water (carbon dioxide dissolved in water, with free carbon dioxide above). So long as the system remains **closed**, the macroscopic properties (e.g. the pressure of CO_2 in the gas and the concentration of the various dissolved materials) will remain constant - the system is in equilibrium. However, on the microscopic scale there is change. Carbon dioxide molecules in the gas will bombard the liquid surface and dissolve; some carbon dioxide molecules in the solution will have sufficient energy to leave the solution and enter the gas phase.

System in equilibrium Go online

A system in equilibrium appears to be unchanging as far as an outside observer is concerned. The bottle of soda water or lemonade shown has carbon dioxide dissolved in the water and also free carbon dioxide above the liquid. The system is **closed** so that nothing can enter or leave the container.

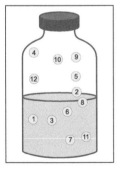
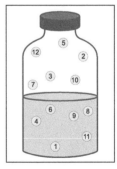

As some carbon dioxide in the gas dissolves, some carbon dioxide in the solution leaves to become gas. So long as the system remains **closed** there is a balance between the rates of the exchange. Notice that the concentrations at equilibrium are not necessarily equal.

At equilibrium these two processes will balance and the number of molecules in the gas and liquid will always be the same, although the individual molecules will not remain static. This state is achieved by a **dynamic equilibrium** between molecules entering and leaving the liquid, and between carbon dioxide, water and carbonic acid. In other words, the rate at which carbonic acid is formed from CO_2 and water will be balanced by carbonic acid dissociating to form CO_2 and water.

$$CO_2 + H_2O \rightleftharpoons H_2CO_3$$

Hydrogen Iodide equilibrium

Go online

Look at the following figure of the reaction between hydrogen and iodine to produce hydrogen iodide in the *forward reaction*.

$$H_2 + I_2 \rightarrow 2HI$$

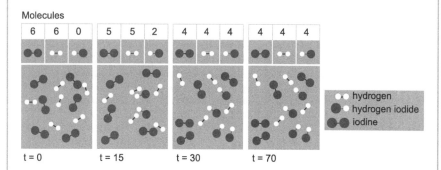

Q1: At t = 0 there are six molecules of H_2, six of I_2 and none of HI. Count the number of molecules of H_2, I_2, and HI after time, t = 15, t = 30 and t = 70. What do you notice about them?

After t = 30 there is no further change in the numbers of reactant and product molecules. Therefore this system has reached equilibrium.

The following figure shows only hydrogen iodide molecules of the *reverse reaction*:

$$2HI \rightarrow H_2 + I_2$$

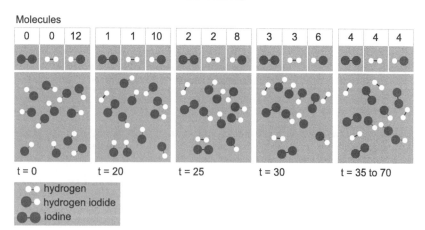

In this case, two molecules of hydrogen iodide react to form hydrogen and iodine.

Q2: What do you notice about the number of H_I, H_2 and I_2 molecules after time, t = 35 and t = 70?

....

Q3: But how do these numbers at equilibrium compare with the previous reaction starting with hydrogen and iodine?

At equilibrium, the rate of production of HI from H_2 and I_2 equals the rate of production of H_2 and I_2 from HI, therefore the overall composition will not change. This process is generally shown by the use of reversible arrows.

$$H_2 + I_2 \rightleftharpoons 2HI$$

The graph shows the progress of the reaction starting with $H_2 + I_2$ to produce an equilibrium mixture of reactants and products.

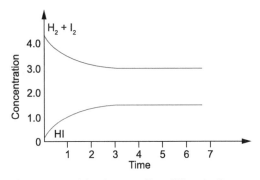

The next graph shows the progress of the decomposition of HI under the same conditions.

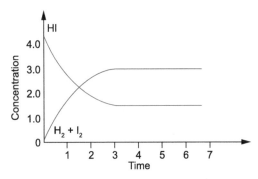

You will notice that the same equilibrium mixture is obtained, whether you start with hydrogen and iodine or with hydrogen iodide.

Consider a simple case of physical equilibrium. A boulder at the top of a hill will remain there until disturbed - it is in a state of equilibrium. When pushed, however, it will readily move into the valley where it will remain even when displaced slightly. The boulder's unstable equilibrium position, shown in the following figure, has moved to a stable equilibrium. This state usually represents a minimum energy state (in this case, the lowest gravitational potential energy).

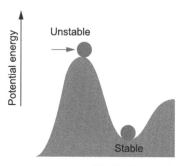

Equilibrium position

Key point

- A chemical reaction is at equilibrium when the composition of the reactants and products remains constant indefinitely.
- This state occurs when the rates of the forward and reverse reactions are equal.
- The same equilibrium mixture is obtained whether you start with reactants or products.

1.2 Equilibrium expressions and factors affecting equilibrium

The equilibrium constant is given the symbol K. It is written as K_c when describing the equilibrium in terms of concentration.

$$aA + bB \rightleftharpoons cC + dD$$

In the equation

$$K_c = \frac{[C]^c[D]^d}{[A]^a[B]^b}$$

where [A], [B], [C] and [D] are the equilibrium concentrations of A, B, C and D respectively and a, b, c and d are the stoichiometric coefficients in a balanced chemical reaction.

The balanced equation for the Haber process is $N_2(g) + 3H_2(g) \rightleftharpoons 2NH_3(g)$

and therefore the equilibrium constant expression is

$$K_c = \frac{[NH_3]^2}{[N_2][H_2]^3}$$

Concentration values are usually measured in mol l^{-1}.

TOPIC 1. CHEMICAL EQUILIBRIUM

Since the equilibrium constant is the ratio of concentration of products divided by the concentration of reactants, its actual value gives guidance to the extent of a reaction once it has reached equilibrium. The greater the value of K_c the greater the concentration of products compared to reactants; in other words, the further the reaction has gone to completion.

The explosive reaction between hydrogen and fluorine:

$$H_2 + F_2 \rightleftharpoons 2HF$$

has an equilibrium constant of 1×10^{47}. At equilibrium, negligible amounts of the reactants will remain; almost all will have been converted to hydrogen fluoride.

In contrast, the dissociation of chlorine molecules to atoms:

$$Cl_2 \rightleftharpoons 2Cl$$

has a K_c value of 1×10^{-38} at normal temperatures, indicating a reaction which hardly occurs at all under these circumstances.

In time, all reactions can be considered to reach equilibrium. To simplify matters, the following general assumption is made:

Value of K_c	Extent of reaction
$< 10^{-3}$	Effectively no reaction
10^{-3} to 10^3	Significant quantities of reactants and products at equilibrium
$> 10^3$	Reaction is effectively complete

A note of caution:

The equilibrium constant gives no indication of the rate at which equilibrium is achieved. It indicates only the ratios of products to reactants once this state is reached.

Equilibrium expressions 1

Q4: Write an equilibrium expression for the following reaction.

$$2Fe^{3+}(aq) + 3I^-(aq) \rightleftharpoons 2Fe^{2+}(aq) + I_3^-(aq)$$

....

Q5: Write an equilibrium expression for the following reaction.

$$H_3PO_4(aq) \rightleftharpoons 2H^+(aq) + HPO_4^{2-}(aq)$$

Returning to the reaction:

$$H_2 + I_2 \rightleftharpoons 2HI$$

the equilibrium constant is defined as:

$$K_c = \frac{[HI]^2}{[H_2][I_2]}$$

and at 453°C, it has a value of 50.

Equilibrium reactions

Q6: At 453°C which compound is present in greatest concentration?

a) Hydrogen
b) Iodine
c) Hydrogen iodide
d) All the same concentration

....

Q7: The K_c value for the reaction $PCl_5 \rightleftharpoons PCl_3 + Cl_2$ is 0.021 at 160°C. Which compound is present in greatest concentration at equilibrium?

a) Phosphorus(V) chloride
b) Phosphorus(III) chloride
c) Chlorine
d) All are the same

....

Q8: The following equilibrium constants apply at room temperature (25 °C).

$$Zn(s) + Cu^{2+}(aq) \rightleftharpoons Cu(s) + Zn^{2+}(aq) \; K_c = 2 \times 10^{37}$$
$$Mg(s) + Cu^{2+}(aq) \rightleftharpoons Cu(s) + Mg^{2+}(aq) \; K_c = 6 \times 10^{90}$$
$$Fe(s) + Cu^{2+}(aq) \rightleftharpoons Cu(s) + Fe^{2+}(aq) \; K_c = 3 \times 10^{26}$$

Of the metals Zn, Mg, and Fe, which removes Cu(II) ions from solution most completely?

a) Zn
b) Mg
c) Fe

....

Q9: In which of the following reactions will the equilibrium lie furthest towards products?

a) $N_2O_4(g) \rightleftharpoons 2NO_2(g)$ K_c at 0°C = 159
b) $2SO_2(g) + O_2(g) \rightleftharpoons 2SO_3(g)$ K_c at 856°C = 21.1
c) $N_2O_4(g) \rightleftharpoons 2NO_2(g)$ K_c at 25°C = 14.4
d) $2SO_2(g) + O_2(g) \rightleftharpoons 2SO_3(g)$ K_c at 636°C = 3343

....

TOPIC 1. CHEMICAL EQUILIBRIUM

Q10: From the data in the previous question, what do you notice about the value of K_c for the oxidation of SO_2 at 856°C compared with 636°C?

....

Q11: Would the process to manufacture SO_3 be more productive at:

a) 636°C
b) 856°C

For gaseous reactions partial pressures may be used. Gases inside a closed container each exert a pressure proportional to the number of moles of the particular gas present (for example, if two gases are mixed in equimolar amounts and the total pressure is 1 atmosphere, then the partial pressure of each gas is 0.5 atmosphere).

$$N_2(g) + 3H_2(g) \rightleftharpoons 2NH_3(g)$$

$$K_p = \frac{(pNH_3)^2}{(pN_2)(pH_2)^3}$$

Equilibrium expressions 2

Q12: Write down an appropriate expression for the equilibrium constant for the following reaction:

$$2NOCl(g) \rightleftharpoons 2NO(g) + Cl_2(g)$$

....

Q13: Write down an appropriate expression for the equilibrium constant for the following reaction:

$$2SO_2(g) + O_2(g) \rightleftharpoons 2SO_3(g)$$

The equilibrium constant has no units whatever the concentrations are measured in.

The example of the Haber process is an example of a homogenous equilibrium where all the species are in the same phase. In heterogeneous equilibria the species are in different phases; an example showing this is through heating calcium carbonate in a closed system. The carbon dioxide formed cannot escape, setting up equilibrium.

$$CaCO_3(s) \rightleftharpoons CO_2(g) + CaO(s)$$

In this reaction where solids exist at equilibrium their concentration is taken as being constant and given the value of 1.

So instead of

$$K_c = \frac{[CO_2(g)][CaO(s)]}{[CaCO_3(s)]}$$

the equilibrium expression is written as $K = [CO_2(g)]$.

© HERIOT-WATT UNIVERSITY

This is also true for pure liquids (including water) where their equilibrium concentration is given the value of 1. This is not true however for aqueous solutions.

Changing the concentration or pressure has an effect on the position of equilibrium (Higher Chemistry, Chemistry in society) however, the equilibrium constant K is not affected.

For example, $NH_3(g) + H_2O(l) \rightleftharpoons NH_4^+(aq) + OH^-(aq)$

$$K = \frac{[NH_4^+][OH^-]}{[NH_3]}$$

If more ammonium ions in the form of solid ammonium chloride are added to the equilibrium the position will shift to the left since ammonium ions are present on the right hand side of the equilibrium (follows Le Chatelier's principle). This increases the concentration of the ammonium ions which causes the system to react in order to decrease the concentration to restore equilibrium. This alters the position of equilibrium until the ratio of products to reactants is the same as before re-establishing the value of K.

Changes in pressure only affect reactions involving gases. If the pressure is increased then the position of equilibrium will shift to the side with the fewer number of gaseous moles. This then causes a new equilibrium to be established, but with the same value of K.

Changes in temperature affect the value of the equilibrium constant K as it is temperature dependent.

$$\text{Reactants} \rightleftharpoons \text{Products}$$

If the forward reaction is exothermic an increase in temperature favours an increase in the concentration of the reactants affecting the ratio [products]/[reactants] by decreasing it. The value of K therefore is decreased.

If the forward reaction is endothermic an increase in temperature favours an increase in the concentration of the products increasing the ratio [products]/[reactants] and increasing the value of K.

1.3 Phase equilibria

Partition coefficients

Immiscible liquids do not mix with each other, the liquid with the lesser density floating on the liquid with the greater density.

Two immiscible liquids are cyclohexane and an aqueous solution of potassium iodide. Solid iodine dissolves in both these liquids and when shaken in both these liquids some dissolves in one liquid while the remainder stays dissolved in the other. The iodine partitions itself between the two liquids. Some of the solute dissolved in the lower level starts to move into the upper layer while at the same time solute in the upper layer starts to move to the lower layer. Eventually the rate of movement from the lower layer to the upper level becomes the same as the rate of the movement from the upper layer to the lower layer and a dynamic equilibrium is set up.

$$I_2(aq) \rightleftharpoons I_2(C_6H_{12})$$

$$K = \frac{[I_2(C_6H_{12})]}{[I_2(aq)]}$$

TOPIC 1. CHEMICAL EQUILIBRIUM

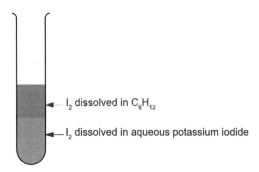

I_2 dissolved in C_6H_{12}

I_2 dissolved in aqueous potassium iodide

Equilibrium constant is known as the partition coefficient and is temperature dependent. It is affected by what solvents/solute are used, but not by adding more solvent or solute.

Solvent extraction

Partition can be used to extract and purify a desired product from a reaction mixture using a separating funnel. This method relies on the product being more soluble in one liquid phase than the other. Caffeine for example is more soluble in dichloromethane than water and this can be used to produce decaffeinated coffee. Due to the harmful nature of dichloromethane caffeine is now extracted using supercritical carbon dioxide, which acts like both a liquid and a gas. It is more efficient to use smaller quantities of the liquid carrying out the extraction a few times rather than using the whole volume at once.

Suppose that for a carboxylic acid partitioning between ether and water:

$$K = \frac{[\text{acid}] \text{ in ether}}{[\text{acid}] \text{ in water}}$$
$$= \frac{[Ae]}{[Aw]}$$
$$= 5$$

If 10 g of acid is dissolved in 100 cm^3 of water and 100 cm^3 of ether is available for extraction, the difference in the quantity extracted by 100 cm^3 of ether used in one extraction of 100 cm^3 or in four extractions of 25 cm^3 can be calculated as follows.

Let v g of acid be extracted with 100 cm^3 ether in a single extraction:

$$K = \frac{[Ae]}{[Aw]} = 5$$
$$\frac{v/100}{(10-v)/100} = 5$$
$$\frac{v}{10-v} = 5$$
$$v = 8.33 \text{ g} \quad \text{i.e. } 8.33 \text{ g is extracted into the ether}$$

When 4 × 25 cm^3 portions of ether are used, the calculation has to be repeated four times.

Let w, x, y and z g of acid be extracted in each successive extraction.

First extraction

$$K = \frac{[Ae]}{[Aw]} = 5$$

$$\frac{w/25}{(10-w)/100} = 5$$

$$4w = 50 - 5w$$

$$w = 5.56 \text{ g}$$

This means that 4.44 g remains in the water to be extracted by the next 25 cm³ of ether.

Second extraction

$$\frac{x/25}{(4.44-x)/100} = 5$$

$$4x = 22.2 - 5x$$

$$x = 2.47 \text{ g}$$

Now 1.97 g remains.

Third extraction

$$\frac{y/25}{(1.97-y)/100} = 5$$

$$4y = 9.85 - 5y$$

$$y = 1.09 \text{ g}$$

Now 0.88 g remains.

Fourth extraction

$$\frac{z/25}{(0.88-z)/100} = 5$$

$$4z = 4.4 - 5z$$

$$z = 0.49 \text{ g}$$

Total amount of carboxylic acid extracted

= w + x + y + z
= 5.56 + 2.47 + 1.09 + 0.49
= 9.61 g

This calculation shows that an extra 1.28 g (9.61 - 8.33 g) of carboxylic acid can be extracted when 4 × 25 cm³ extractions are used rather than 1 × 100 cm³ extraction.

1.3.1 Chromatography

All chromatographic methods involve a mobile phase moving over a stationary phase. Separation occurs because the substances in the mixture have different partition coefficients between the stationary and mobile phases.

Substances present in the initial mixture which partition more strongly into the stationary phase will move more slowly than materials which partition more strongly into the mobile phase.

In paper chromatography the mixture of components to be separated is placed as a small spot close to the bottom of a rectangular piece of absorbent paper (like filter paper). The bottom of the paper is placed in a shallow pool of solvent in a tank. An example of the solvent would be an alcohol. Owing to capillary attraction, the solvent is drawn up the paper, becoming the mobile phase. The solvent front is clearly visible as the chromatography progresses.

When the paper is removed from the solvent, the various components in the initial spot have moved different distances up the paper.

Paper chromatography — Go online

A simple simulation of this process, showing chromatography of blue and black inks, is available on the on-line version of this Topic. The start and final positions are shown in Figure 1.1 and Figure 1.2 respectively.

Figure 1.1: Start of chromatographic analysis

Figure 1.2: End of chromatographic analysis

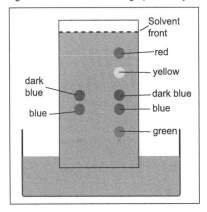

Q14: By observing the chromatography simulation, which ink (blue or black) has the most components?

a) Blue
b) Black
c) Both the same

....

Q15: Which material in the black ink has stayed longest on the stationary phase, and has the lowest solvent/water partition coefficient?

a) Red
b) Yellow
c) Dark Blue
d) Blue
e) Green

....

Q16: The red coloured spot has moved furthest, this would indicate that the red material:

a) has the highest solvent/water partition coefficient.
b) has the lowest solvent/water partition coefficient.
c) has the lowest molecular mass.

....

Q17: The movement of materials on paper chromatography is often described by an R_f value which is the distance travelled by the spot divided by the distance travelled by the solvent front. As long as the conditions of chromatography remain the same, a compound will have a constant R_f value.

Which colour in the black ink could have an R_f value of 0.4?

a) Red
b) Yellow
c) Dark blue
d) Blue
e) Green

....

Q18: Both the blue and dark blue spots from both the original inks have moved similar distances. What might you conclude from this?

This separation depends on the different partition coefficients of the various components. The components are partitioned between the solvent and the water trapped in the paper. Substances which partition mainly into the solvent mobile phase will move further up the paper than substances which partition more strongly onto the stationary phase.

Although paper chromatography is still used today, it has been largely replaced by thin layer chromatography (TLC). In this method, a support of glass or aluminium is coated, usually with a thin layer of silica or cellulose. The processing is identical to that described previously, but TLC allows a more rapid separation (which prevents the spots spreading too far) and makes detection of the spots easier. Most materials are not coloured, but can still be chromatographed.

The invisible spots on paper or thin layer chromatography are revealed by use of a locating reagent. These react with the compounds in the spots to produce a coloured derivative. For example, ninhydrin solution can be sprayed onto chromatograms to reveal amino-acids.

In another TLC detection system, the silica stationary phase is mixed with a fluorescent dye, so that at the end of the process, viewing the plate under an ultraviolet lamp will cause the background to glow (often an eerie green) **except** where there are spots, which appear black.

In gas-liquid chromatography (**GLC**) the stationary phase is a high boiling point liquid held on an inert, finely-powdered support material, and the mobile phase is a gas (often called the carrier gas). The stationary phase is packed into a tubular column usually of glass or metal, with a length of 1 to 3 metres and internal diameter about 2 mm. One end of the column is connected to a gas supply (often nitrogen or helium) via a device which enables a small volume of liquid sample (containing the mixture to be analysed) to be injected into the gas stream. The other end of the column is connected to a device which can detect the presence of compounds in the gas stream. The column is housed in an oven to enable the temperature to be controlled throughout the chromatographic analysis. One reason for this is that the materials to be analysed must be gases during the analysis, so that gas-liquid chromatography is often carried out at elevated temperatures.

A mixture of the material to be analysed is injected into the gas stream at zero time. The individual components travel through the packed column at rates which depend on their partition coefficients between the liquid stationary phase and the gaseous mobile phase. The detector is set to measure some change in the carrier gas that signals the presence of material coming from the end of the column. Some detectors measure the thermal conductivity of the gas, others burn the material from the column in a hydrogen-air flame and measure the presence of ions in the flame. The signal from the detector is recorded and plotted against time to give a series of peaks each with an individual retention time.

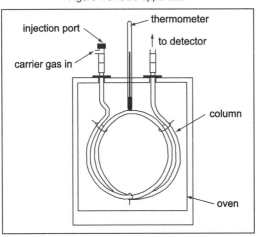

Figure 1.3: GLC apparatus

1.4 Acid/base equilibria

In acids the concentration of hydrogen ions (H^+) is greater than the concentration of hydrogen ions in water. The concentration of hydroxide ions (OH^-) is greater in alkalis than the concentration of hydroxide ions in water.

A hydrogen ion is basically a proton (hydrogen atom that has lost an electron) and only exists when surrounded by water molecules in an aqueous solution. These are known as hydronium ions and are written as $H_3O^+(aq)$ but are often shortened to just $H^+(aq)$.

An acid therefore is a proton donor and a base is a proton acceptor. This definition was put forward by Bronsted and Lowry in 1932. When an acid donates a proton the species left is called the **conjugate base**.

$$HA \rightleftharpoons H_3O^+ + A^-$$
$$\text{Acid} \qquad\qquad\qquad \text{Conjugate base}$$

When a base accepts a proton the species formed is a **conjugate acid**.

$$B + H_3O^+ \rightleftharpoons BH^+$$
$$\qquad \text{Base} \qquad\qquad \text{Conjugate acid}$$

Acid	Base	Conjugate acid	Conjugate base
HNO_3	H_2O	H_3O^+	NO_3^-
$HCOOH$	H_2O	H_3O^+	$HCOO^-$
H_2O	NH_3	NH_4^+	OH^-
H_2O	F^-	HF	OH^-

1.4.1 Ionisation of water

From the table we can see that water can act as both an acid and a base. Therefore water can be called **amphoteric**, i.e. can behave as either a base or an acid.

In water and aqueous solutions there is an equilibrium between the water molecules, and hydronium (hydrogen) and hydroxide ions. Water acts both as a proton donor (acid) forming a conjugate base and the proton acceptor (base) forming a conjugate acid.

This (dissociation) ionisation of water can be represented by:

$$H_2O(\ell) \text{ (acid)} + H_2O(\ell) \text{ (base)} \rightleftharpoons H_3O^+(aq) + OH^-(aq)$$
$$\qquad\qquad\qquad\qquad\qquad\qquad \text{Conjugate acid} \qquad \text{Conjugate base}$$

The equilibrium constant K is defined by $K = [H_3O^+(aq)][OH^-(aq)]$ or simply $[H_3O^+][OH^-]$ and is known as the **ionic product of water** and given the symbol K_w. This is temperature dependent and the value is approximately 1×10^{-14} at 25°C.

TOPIC 1. CHEMICAL EQUILIBRIUM

1.4.2 pH scale

The concentration of H_3O^+ (H^+) and OH^- in pure water is very small. At 25°C the concentration of H_3O^+ (and OH^-) is 0.0000001 in units of moles per litre. The square brackets around the symbol H_3O^+, i.e. $[H_3O^+]$, is the *concentration in moles per litre*. In scientific notation this is 1.0×10^{-7} mol ℓ^{-1}.

Rather than use these very small fractions, chemists convert the H_3O^+ concentrations to a new scale - the pH scale - which uses small positive numbers. The way to do this is to define the pH as:

$$pH = -\log_{10}[H_3O^+]$$

$[H_3O^+]$	10^{-2}	10^{-4}	10^{-6}	**10^{-7}**	10^{-8}	10^{-10}	10^{-12}	10^{-14}
$\log_{10}[H_3O^+]$	-2	-4	-6	**-7**	-8	-10	-12	-14
$-\log_{10}[H_3O^+]$	+2	+4	+6	**+7**	+8	+10	+12	+14
pH	2	4	6	**7**	8	10	12	14

> **Key point**
>
> pH values are not always whole numbers.

Acids pH values lower than 7 $[H_3O^+] > [OH^-]$

Bases pH values higher than 7 $[H_3O^+] < [OH^-]$

Neutral solutions pH values = 7 $[H_3O^+] = [OH^-]$

The relationship between pH and hydrogen ion concentration is given by the equation:

$$pH = -\log_{10}[H_3O^+]$$

This can be used to calculate the pH of strong acids and alkalis (see later for information on these). The pH can be used to calculate the hydronium ion concentration using the equation: $[H_3O^+] = 10^{-pH}$

> **Example : Calculate the pH**
>
> Calculate the pH of 0.21 mol l^{-1} HCl (aq)
>
> $[H_3O^+] = 0.21$ mol l^{-1}
>
> $pH = -\log_{10}(0.21) = 0.68$
>
> You should also be able to calculate concentrations from the pH value.

Example : $[H_3O^+]$ from a pH value

Calculate $[H_3O^+]$ from a pH value of 11.6

$11.6 = -\log_{10}[H_3O^+]$

$[H_3O^+] = 10^{-11.6}$

$[H_3O^+] = 2.51 \times 10^{-12}$ mol l^{-1}

The pH of neutral solutions can be calculated to be 7 as follows:

$$K_w = [H_3O^+][OH^-] = 1 \times 10^{-14}$$

As in neutral solutions $[H_3O^+] = [OH^-]$ then both the concentration of H_3O^+ and OH^- are 1×10^{-7}.

$$pH = -\log_{10}(1 \times 10^{-7}) = 7$$

This pH of 7 for neutral solutions only applies at 25°C. The pH will decrease as temperature increases due to the fact that the ionisation of water is an endothermic process, therefore the value of K_w will increase with increasing temperature. See equilibrium constants changing with temperature in the section 'Equilibrium expressions and factors affecting equilibrium' earlier in this topic.

Key point

The relationship between pH and the hydrogen ion concentration is given by

$$pH = -\log_{10}[H_3O^+]$$

This relationship can be used to calculate the pH for strong acids and alkalis given the molar concentrations of either H_3O^+ or OH^- ions.

Calculating pH   Go online

Practice calculating pH values for strong acids and alkalis and at calculating $[H_3O^+]$ and $[OH^-]$ from pH values.

Examples

1. pH from $[H_3O^+]$ (1)

Calculate the pH of a 0.02 mol l^{-1} solution of hydrochloric acid.

$$[H_3O^+] = 0.02 \text{ mol l}^{-1}$$
$$= 2 \times 10^{-2} \text{ mol l}^{-1}$$
$$-\log[H_3O^+] = 1.7$$
$$pH = 1.7$$

It is always useful to check your answer by estimating values.

0.02 lies between 0.01 and 0.10, i.e. between 10^{-2} and 10^{-1}.

So the pH must lie between 2 and 1.

TOPIC 1. CHEMICAL EQUILIBRIUM

2. pH from [H₃O⁺] (2)

Calculate the pH of a 0.001 mol l⁻¹ solution of sulfuric acid.

Since there are 2 moles of H_3O^+ ions present per mole of H_2SO_4 then:

$$[H_3O^+] = 0.002 \text{ mol l}^{-1}$$
$$= 2 \times 10^{-3} \text{ mol l}^{-1}$$
$$-\log[H_3O^+] = 2.7$$
$$pH = 2.7$$

3. pH from [OH⁻]

Calculate the pH of a solution of 0.006 mol l⁻¹ sodium hydroxide.

$$[OH^-] = 0.006$$
$$= 6 \times 10^{-3} \text{ mol l}^{-1}$$
$$K_w = [H_3O^+][OH^-]$$
$$[H_3O^+] = \frac{K_w}{[OH^-]}$$
$$= \frac{10^{-14}}{6 \times 10^{-3}}$$
$$= 1.667 \times 10^{-12}$$

Therefore:

$$pH = -\log(1.667) - \log(10^{-12})$$
$$= -0.22 + 12$$
$$= \underline{11.78}$$

Check $[H_3O^+]$ lies between 10^{-11} and 10^{-12}, i.e pH lies between 11 and 12.

Q19: Calculate the pH of a solution that has a H_3O^+(aq) concentration of 5×10^{-3} mol ℓ^{-1}. Give your answer to 2 decimal places.

Q20: Calculate the pH of a solution that has a H_3O^+(aq) concentration of 8×10^{-6} mol ℓ^{-1}. Give your answer to 2 decimal places.

Q21: Calculate the pH of a solution that has a OH^-(aq) concentration of 6.3×10^{-2} mol ℓ^{-1}. Give your answer to 2 decimal places.

© HERIOT-WATT UNIVERSITY

Q22: Calculate the pH of a solution that has a OH⁻(aq) concentration of 2.9×10^{-4} mol ℓ^{-1}. Give your answer to 2 decimal places.

Using the same relationship $\left(\text{pH} = -\log_{10}[H_3O^+]\right)$, the concentration of H_3O^+ ions and OH⁻ ions can be calculated from the pH of the solution.

Example : Concentrations from pH

Calculate the concentration of H_3O^+ ions and OH⁻ ions in a solution of pH 3.6. Give your answer to three significant figures.

Step 1

$$\begin{aligned}
\text{pH} &= -\log[H_3O^+] \\
3.6 &= -\log[H_3O^+] \\
\log[H_3O^+] &= -3.6 \\
[H_3O^+] &= \text{antilog}(-3.6) \\
&= 10^{-3.6} \\
&= 0.000251 \\
&= \underline{\underline{2.51 \times 10^{-4}}} \text{ mol l}^{-1}
\end{aligned}$$

Note: If $\log[H_3O^+] = x$ then $[H_3O^+] = 10^x$. Calculators vary slightly in the way in which they antilog numbers. One way is as follows:

$$\begin{aligned}
\text{If } \log[H_3O^+] &= x \\
\text{then } [H_3O^+] &= 10^x
\end{aligned}$$

Press the 10^x key, then type the value (-3.6), followed by '='.

If this does not work with your calculator, see your tutor.

Step 2

$$\begin{aligned}
K_w &= [H_3O^+][OH^-] \\
[OH^-] &= \frac{K_w}{[H_3O^+]} \\
&= \frac{10^{-14}}{2.51 \times 10^{-4}} \\
&= \underline{\underline{3.98 \times 10^{-11}}} \text{ mol l}^{-1}
\end{aligned}$$

An alternative to step 2 uses the equation pH + pOH = 14.

TOPIC 1. CHEMICAL EQUILIBRIUM

Alternative Step 2:

$$pH + pOH = 14$$
$$pOH = 14 - 3.6 = 10.4$$
$$-\log[OH^-] = 10.4$$
$$[OH^-] = 10^{-10.4}$$
$$= \underline{3.98 \times 10^{-11}} \text{ mol l}^{-1}$$

Q23: Calculate the concentration of H_3O^+(aq) ions and OH^-(aq) ions in a solution of pH 2.3

....

Q24: Calculate the concentration of H_3O^+(aq) ions and OH^-(aq) ions in a solution of pH 5.6

....

Q25: Calculate the concentration of H_3O^+(aq) ions and OH^-(aq) ions in a solution of pH 11.4

....

Q26: Calculate the concentration of H_3O^+(aq) ions and OH^-(aq) ions in a solution of pH 1.9

1.4.3 Strong/weak acids and bases

Strong acids and bases are those that completely ionise (dissociate) when dissolved in water. Examples of strong acids include hydrochloric acid, sulfuric acid and nitric acid.

$$HCl(g) + H_2O(l) \rightarrow H_3O^+(aq) + Cl^-(aq)$$

Hydrochloric acid completely dissociates into hydrogen ions and chloride ions when dissolved in water.

Examples of strong bases include sodium hydroxide, potassium hydroxide and calcium hydroxide.

$$NaOH(s) \rightarrow Na^+(aq) + OH^-(aq)$$

Sodium hydroxide completely dissociates into sodium ions and hydroxide ions when dissolved in water.

For strong monoprotic acids (acids with only one hydrogen ion in their formula, e.g. HCl), the hydrogen ion concentration is the same as the original concentration of the acid as all the acid molecules have dissociated into ions. For strong diprotic acids (acids with two hydrogen ions in their formula, e.g. H_2SO_4), the hydrogen ion concentration will be double that of the original acid concentration since all the hydrogen ions will be released in solution.

Weak acids are only partially ionised (dissociated) when dissolved in water and an equilibrium is set up which lies to the left. Approximately only 1% of the acid molecules are dissociated and therefore the hydrogen ion concentration in solution will be much less than the concentration of the acid. Examples of weak acids include ethanoic acid (all carboxylic acids), carbonic acid and sulfurous acid.

$$CH_3COOH(l) + H_2O(l) \rightleftharpoons CH_3COO^-(aq) + H_3O^+(aq)$$

This is also true for weak bases. Examples of weak bases include ammonia (NH_3) and amines (CH_3NH_2).

$$NH_3(g) + H_2O(l) \rightleftharpoons NH_4^+(aq) + OH^-(aq)$$

We can compare conductivity, pH, rate of reaction and volume to neutralise alkali of strong and weak acids.

Property	Strong acid	Weak acid
Conductivity	Higher	Lower
Rate of Reaction	Faster	Slower
pH	Lower	Higher
Volume to neutralise acid	Same	Same

The differences in conductivity, pH and rate of reaction can be attributed to the fact that strong acids have a much higher number of hydrogen ions in solution than weak acids of the same concentration. However, the volume of alkali required to neutralise a strong and a weak acid of the same concentration is the same. The hydroxide ions in the alkali react with all of the available hydrogen ions in solution. However, in a weak acid, this removes hydrogen ions from the equilibrium and causes the acid molecules to release more hydrogen ions. This continues until all the acid molecules have dissociated, i.e. until the acid is neutralised. The volume of alkali required therefore depends only on the concentration of the acid and not on the strength of the acid.

We can compare conductivity, pH and volume to neutralise acid of weak and strong bases.

Property	Strong base	Weak base
Conductivity	Higher	Lower
pH	Higher	Lower
Volume to neutralise alkali	Same	Same

To calculate the pH of weak acids we need to use the formula $pH = \frac{1}{2} pK_a - \frac{1}{2} \log c$

$$HA \rightleftharpoons H_3O^+ + A^-$$

At equilibrium the $[H_3O^+] = [A^-]$ and since the equilibrium lies very far to the left hand side, $[HA]$ at equilibrium is approximately the same as the original concentration of the acid, c. This can therefore be written as $K_a = [H_3O^+]^2/c$.

Therefore $[H_3O^+]^2 = K_a c$ and $[H_3O^+] = \sqrt{K_a c}$.

Since $pH = -\log[H_3O^+]$, it follows that $pH = -\log \sqrt{K_a c}$.

As $-\log K_a = pK_a$, $pH = -\frac{1}{2}\log K_a - \frac{1}{2}\log c$.

The equation then becomes $pH = \frac{1}{2}pK_a - \frac{1}{2}\log c$.

The larger the value of K_a the stronger the acid (or vice versa) and the smaller the pK_a the stronger the acid.

TOPIC 1. CHEMICAL EQUILIBRIUM

If we use the equation as a general formula for a weak acid dissociating we can write the equilibrium constant of the acid as

$$K_a = \frac{[H_3O^+][A^-]}{[HA]}$$

$[H_3O^+] = \sqrt{K_a c}$ where c is the concentration of the acid.

The dissociation constant of a weak acid can be represented by $pK_a = -\log K_a$ which is often more convenient to use than K_a and relates to the equation for pH of a weak acid.

In weak bases the dissociation constant is written as:

$$K_b = \frac{[NH_4^+][OH^-]}{[NH_3]}$$

The ammonium ion formed is a weak acid and will dissociate as follows:

$$NH_4^+(aq) + H_2O(l) \rightleftharpoons NH_3(aq) + H_3O^+(aq)$$

This time the H_2O is the base, conjugate base is NH_3 and the conjugate acid H_3O^+.

Dissociation constant K_a for the ammonium ion is

$$K_a = \frac{[NH_3][H_3O^+]}{[NH_4^+]}$$

The greater the numerical value of K_a for a weak acid the stronger it is. The smaller the pK_a value, the stronger the acid. K_a and pK_a values are given in page 13 of the Chemistry Data Booklet (http://bit.ly/2qPDbBg).

Strong/weak acids and bases

Q27: Calculate the pH of a 0.1 mol ℓ^{-1} solution of ethanoic acid. Answer to two decimal places.

1.4.4 Salts

Salts are most simply defined as one of the products of the neutralisation of an acid by a base. More accurately, a salt is formed when the hydrogen ions of an acid are replaced by metal ions or ammonium ions.

The first part of the name of a salt identifies which alkali/base was used to make the salt. Sodium salts are generally made using sodium hydroxide and potassium salts made using potassium hydroxide as the alkali.

The second part of the name of a salt identifies the acid used to make the salt. Chloride salts are made from hydrochloric acid, nitrate salts nitric acid and sulfate salts sulfuric acid.

Parent acid and base

Q28: Identify the parent acid and base used to form magnesium nitrate.
....

Q29: Identify the parent acid and base used to form potassium bromide.
....

Q30: Identify the parent acid and base used to form sodium ethanoate.
....

Q31: Identify the parent acid and base used to form calcium sulfite. (Care - note the different name ending.)

Some salt solutions are neutral but not all. The pH of a salt solution depends on the relative strengths of the parent acid and parent base. You can imagine the acid trying to pull the pH towards the acidic side but being opposed by the base pulling in the opposite direction, as in a tug of war. Whichever is stronger will pull the pH towards its end of the scale.

Figure 1.4: pH scale and universal indicator colour

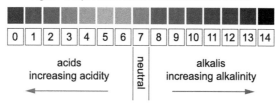

- If the parent acid is stronger than the parent base, the pH of the salt solution will be less than 7, i.e. acidic.
- If the parent base is stronger than the parent acid, the pH of the salt solution will be greater than 7, i.e. alkaline.
- If the acid and base are equally strong, the salt solution will be neutral (pH 7).

For example potassium ethanoate will have been made using the parent acid ethanoic acid (weak acid) and the alkali potassium hydroxide (strong base) and therefore the pH of potassium ethanoate will be more than 7 and alkaline.

Salts

Q32: Which of the following is likely to be the pH of a solution of sodium ethanoate?

a) 5
b) 7
c) 9

....

TOPIC 1. CHEMICAL EQUILIBRIUM

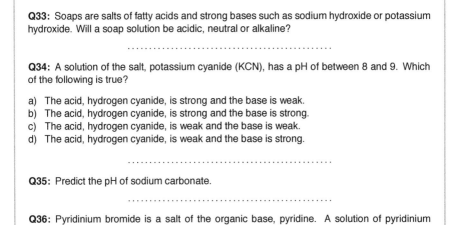

Q33: Soaps are salts of fatty acids and strong bases such as sodium hydroxide or potassium hydroxide. Will a soap solution be acidic, neutral or alkaline?

....

Q34: A solution of the salt, potassium cyanide (KCN), has a pH of between 8 and 9. Which of the following is true?
a) The acid, hydrogen cyanide, is strong and the base is weak.
b) The acid, hydrogen cyanide, is strong and the base is strong.
c) The acid, hydrogen cyanide, is weak and the base is weak.
d) The acid, hydrogen cyanide, is weak and the base is strong.

....

Q35: Predict the pH of sodium carbonate.

....

Q36: Pyridinium bromide is a salt of the organic base, pyridine. A solution of pyridinium bromide has a pH of between 5 and 6. explain whether pyridine is a strong base or a weak base.

In order to explain the pH of a salt solution, we need to consider the equilibria involved, in particular the effect on the water equilibrium. If this equilibrium is not affected, the pH will remain the same as in pure water (pH 7). Any change in the proportions of $H_3O^+(aq)$ (H^+) and $OH^-(aq)$ ions will cause a change in pH. Three types of salt will be considered in turn.

1. salts of strong acids and strong bases;
2. salts of strong bases and weak acids;
3. salts of strong acids and weak bases.

Strong acid and strong base

Potassium nitrate solution has a pH of 7.

Potassium nitrate is the salt of a strong acid and a strong base (nitric acid and potassium hydroxide respectively). When the salt dissolves in water, the solution contains a high concentration of $K^+(aq)$ ions and $NO_3^-(aq)$ ions (from the salt) and a very low concentration of $H_3O^+(aq)$ ions and $OH^-(aq)$ ions (from the dissociation of water).

$$H_2O\,(l) \rightleftharpoons H_3O^+(aq) + OH^-(aq)$$

Since potassium hydroxide is strong, $K^+(aq)$ ions will not react with $OH^-(aq)$ ions. Since nitric acid is strong, $NO_3^-(aq)$ ions will not react with $H_3O^+(aq)$ ions. The water equilibrium will not be affected. The $[H_3O^+(aq)]$ and $[OH^-(aq)]$ will remain the same as in pure water (both equal to 10^{-7} mol l^{-1}) and the solution will be neutral.

Strong base and weak acid

Sodium ethanoate solution has a pH of 9

Sodium ethanoate is the salt of a strong base and a weak acid (sodium hydroxide and ethanoic acid respectively). When the salt dissolves in water, the solution contains ethanoate ions ($CH_3COO^-(aq)$) and sodium ions ($Na^+(aq)$) from the salt and $H_3O^+(aq)$ ions and $OH^-(aq)$ ions (from the dissociation of water). Sodium hydroxide is a strong base (100% ionised). So, if a sodium ion meets a hydroxide ion in the solution, they will not combine. However, ethanoic acid is a weak acid.

$$CH_3COOH(aq) \rightleftharpoons CH_3COO^-(aq) + H_3O^+(aq)$$

In ethanoic acid solution, this equilibrium lies well to the left. There are lots of molecules and few ions. Remember that the same equilibrium position is reached from either direction, i.e it does not matter whether you start with 100% molecules or 100% ions. As soon as sodium ethanoate dissolves in water, the concentration of ethanoate ions is much higher than the equilibrium concentration. So ethanoate ions combine with $H_3O^+(aq)$ ions (from water) to form ethanoic acid molecules until equilibrium is established. The removal of $H_3O^+(aq)$ ions disrupts the water equilibrium.

$$H_2O\ (l) \rightleftharpoons H_3O^+(aq) + OH^-(aq)$$

More water molecules dissociate in an attempt to replace the removed $H_3O^+(aq)$ ions until a new equilibrium is established. The result is an excess of $OH^-(aq)$ ions and an alkaline pH.

Strong acid and weak base

Ammonium Chloride has a pH of 4.

Ammonium Chloride is the salt of a weak base and a strong acid (ammonia and hydrochloric acid respectively). When the salt dissolves in water the solution contains ammonium ions ($NH_4^+(aq)$) and chloride ions ($Cl^-(aq)$) from the salt and $H_3O^+(aq)$ ions and $OH^-(aq)$ ions from the dissociation of water. Hydrochloric acid is a strong acid (100% ionised) so if a chloride ion meets a hydrogen ion in the solution they will not combine. However ammonia is a weak base:

$$NH_3(aq) \rightleftharpoons NH_4^+(aq) + OH^-(aq)$$

In ammonia solution, this equilibrium lies well to the left. There are lots of molecules and few ions. Remember that the same equilibrium position is reached from either direction, i.e it does not matter whether you start with 100% molecules or 100% ions. As soon as ammonium chloride dissolves in water, the concentration of ammonium ions is much higher than the equilibrium concentration. So ammonium ions combine with $OH^-(aq)$ ions (from water) to form ammonia molecules until equilibrium is established. The removal of $OH^-(aq)$ ions disrupts the water equilibrium.

$$H_2O\ (l) \rightleftharpoons H_3O^+(aq) + OH^-(aq)$$

More water molecules dissociate in an attempt to replace the removed $OH^-(aq)$ ions until a new equilibrium is established. The result is an excess of $H_3O^+(aq)$ ions and an acidic pH.

1.5 Indicators and buffers

Indicators

Acid/base indicators (or simply indicators) are weak acids which change colour depending on the pH of the solution.

HIn can be used as a general formula for an indicator and its dissociation can be represented by this equation:

$$HIn(aq) + H_2O(l) \rightleftharpoons H_3O^+(aq) + In^-(aq)$$

For a good indicator, the undissociated acid, HIn, will have a distinctly different colour from its conjugate base, In-.

The acid dissociation constant for an indicator HIn is given the symbol K_{In} and is represented by:

$$K_{In} = \frac{[H_3O^+][In^-]}{[HIn]}$$

Taking the negative log of both sides gives:

$$-\log K_{In} = -\log[H_3O^+] - \log[In^-]/[HIn]$$
$$\text{Since } pK_{In} = -\log K_{In}$$
$$\text{And } pH = -\log[H_3O^+]$$
$$\text{Then } pK_{In} = pH - \log[In^-]/[HIn]$$
$$pH = pK_{In} + \log[In^-]/[HIn]$$

The pH of the solution is determined by the pK_{In} of the indicator and the ratio of **[In-]** to **[HIn]**. Since these are different colours, the ratio of **[In-]** to **[HIn]** determines the overall colour of the solution. For a given indicator, the overall colour of the solution is dependent on the pH of the solution.

[HIn] and [In-] need to differ by a factor of 10 to distinguish the colour change. The pH range for the colour change is estimated by $pK_{In} \pm 1$.

1.5.1 pH titrations

When an acid is gradually neutralised by a base, the change in pH can be monitored using a pH meter. The results can be used to produce a pH titration curve from which the **equivalence point** can be identified.

> **Key point**
>
> During a titration involving a strong acid and base, there is a very rapid change in pH around the equivalence point.

pH titration

In the following figure, 50cm³ hydrochloric acid of concentration 0.1 mol ℓ^{-1} is being neutralised by sodium hydroxide solution.

Use the graph shown as Figure 1.5, to answer the questions.

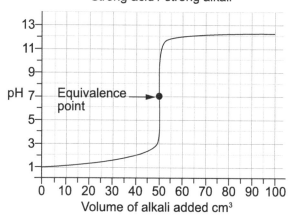

Figure 1.5: Titration curve
Strong acid / strong alkali

Q37: What is the pH at the equivalence point?

Q38: What is meant by the equivalence point?

Q39: Use the information in the graph to calculate the concentration (in mol ℓ^{-1}) of the sodium hydroxide solution.

Q40: Between 49.9 cm³ and 50.0 cm³, only 0.1 cm³ of alkali was added. What was the change in pH for this addition?

Q41: Which of these statements is true?

a) The alkali is more concentrated than the acid.
b) The pH rises rapidly at the beginning.
c) The alkali is less concentrated than the acid.
d) The pH changes rapidly only around the equivalence point.

Titration curves

Figure 1.6: Titration curves

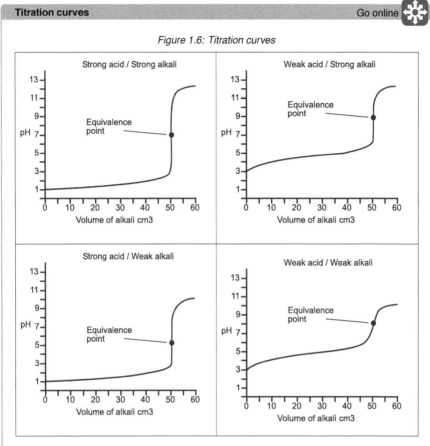

Figure 1.6 shows pH titration curves produced by the different combinations of strong and weak acids and alkalis. Look closely at the equivalence points in each graph.

Q42: Which combination has an equivalence point at pH 7?

a) Strong acid/strong alkali
b) Strong acid/weak alkali
c) Weak acid/strong alkali
d) Weak acid/weak alkali

....

Q43: What is the pH of the salt formed from a strong acid and a strong alkali?

a) 1
b) 5
c) 7
d) 9

....

Q44: Which of these is the most likely pH of the salt formed from a strong acid and a weak alkali?

a) 1
b) 5
c) 7
d) 9

....

Q45: Which of these is the most likely pH of the salt formed from a weak acid and a strong alkali?

a) 1
b) 5
c) 7
d) 9

....

Q46: Using the graphs in Figure 1.6 and your previous answers, write a general statement about the pH at the equivalence point in an acid/alkali titration. Then display the answer.

Since the equivalence points occur at different pH for different combinations, different indicators will be required in each case.

Choosing indicators Go online

There are four different possible combinations of acid and alkali, shown in the following table.

A	strong acid/ strong alkali
B	strong acid/ weak alkali
C	weak acid/ strong alkali
D	weak acid/ weak alkali

Look at the titration curves for these four different combinations of acids and bases, then answer the questions.

TOPIC 1. CHEMICAL EQUILIBRIUM

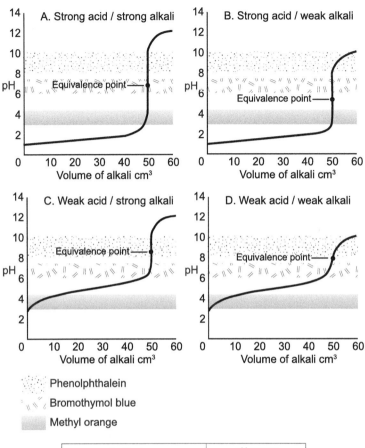

Indicator	pH range
phenolphthalein	8.2 - 10.0
bromothymol blue	6.0 - 7.6
methyl orange	3.0 - 4.4

Other indicators and their pH ranges are given in page 20 of the Chemistry Data Booklet (http://bit.ly/2qPDbBg).

Set A: strong acid and strong base

For each indicator, decide whether or not it is suitable for this titration

Q47: Phenolphthalein

a) Suitable
b) Unsuitable

....

Q48: Bromothymol blue

a) Suitable
b) Unsuitable

....

Q49: Methyl orange

a) Suitable
b) Unsuitable

....

Q50: Which would be the best indicator for this titration?

a) Phenolphthalein
b) Bromothymol blue
c) Methyl orange
d) None of these

....

Q51: Explain your choice.

Set B: Strong acid and weak base

For each indicator, decide whether or not it is suitable for this titration

Q52: Phenolphthalein

a) Suitable
b) Unsuitable

....

Q53: Bromothymol blue

a) Suitable
b) Unsuitable

....

TOPIC 1. CHEMICAL EQUILIBRIUM

Q54: methyl orange

a) Suitable
b) Unsuitable

....

Q55: Which would be the best indicator for this titration?

a) Phenolphthalein
b) Bromothymol blue
c) Methyl orange
d) Either methyl orange or bromothymol blue

....

Q56: Explain your choice.

Set C: Weak acid and strong base

For each indicator, decide whether or not it is suitable for this titration

Q57: Phenolphthalein

a) Suitable
b) Unsuitable

....

Q58: Bromothymol blue

a) Suitable
b) Unsuitable

....

Q59: Methyl orange

a) Suitable
b) Unsuitable

....

Q60: Which would be the best indicator for this titration?

a) Phenolphthalein
b) Bromothymol blue
c) Methyl orange
d) None of these

....

Q61: Explain your choice.

Set D: Weak acid and weak base

For each indicator, decide whether or not it is suitable for this titration

Q62: Phenolphthalein

a) Suitable
b) Unsuitable

....

Q63: Bromothymol blue

a) Suitable
b) Unsuitable

....

Q64: Methyl orange

a) Suitable
b) Unsuitable

....

Q65: Which would be the best indicator for this titration?

a) Phenolphthalein
b) Bromothymol blue
c) Methyl orange
d) None of these

....

Q66: Explain your choice.

1.5.2 Buffer solutions

Small changes in pH can have a surprisingly large effect on a system. For example, adding a small volume of lemon juice or vinegar to milk changes the protein structure and curdling occurs. Many processes, particularly in living systems, have to take place within a precise pH range. Should the pH of blood move 0.5 units outside the range shown in the following table (Table 1.1), the person would become unconscious and die. Evolution has devised buffer solutions to prevent such changes in pH in the body. A **buffer solution** is one in which the pH remains approximately constant when small amounts of acid or base are added.

Table 1.1: pH of body fluids

TOPIC 1. CHEMICAL EQUILIBRIUM

Fluid	pH range
blood	7.35 - 7.45
saliva	6.4 - 6.8
tears	7.4
urine	4.8 - 7.5
stomach juices	1.6 - 1.8

Biological systems work within precise ranges which buffers keep fairly constant. Why do you think urine can have such a wide range?

Manufacturing systems also require precise control of pH and buffers are used in electroplating, photographic work and dye manufacture. Some examples are shown.

Many pharmacy products try to match their pH to the pH of the body tissue.

Electroplating industries need pH control over their plating solutions.

Buffer solutions are of two types:

- an acid buffer consists of a solution of a weak acid and one of its salts
- a basic buffer consists of a solution of a weak base and one of its salts

If a buffer is to stabilise pH, it must be able to absorb extra acid or alkali if these are encountered.

TOPIC 1. CHEMICAL EQUILIBRIUM

1.5.2.1 Acid buffers

An acid buffer consists of a weak acid represented as **HA**. It will be slightly dissociated. Large reserves of **HA** molecules are present in the buffer.

Figure 1.7: Dissociation of a weak acid

$$HA \rightleftharpoons H_3O^+ + A^- \quad \text{at equilibrum}$$

The weak acid salt **MA** also present will be completely dissociated. Large reserves of the **A⁻** ion are present in the buffer. This is the conjugate base.

Figure 1.8: Dissociation of a salt of a weak acid

$$MA \longrightarrow M^+ + A^- \quad \text{complete dissociation}$$

How does the buffer work? If acid is added to the mixture the large reserve of **A⁻** ions will trap the extra hydrogen ions and convert them to the weak acid. This stabilises the pH. If alkali is added the large reserve of **HA** molecules will convert the extra **OH⁻** to water. This again stabilises the pH.

A typical example of an acid buffer solution would be ethanoic acid and sodium ethanoate. The ethanoic acid is only partly dissociated. The sodium ethanoate salt completely dissociates and provides the conjugate base.

Figure 1.9: Equilibria in acid buffer solutions

$$CH_3COOH \rightleftharpoons CH_3COO^- + H_3O^+$$
(HA reserve)

$$CH_3COONa \longrightarrow CH_3COO^- + Na^+$$
(A⁻ reserve)

The stable pH of the buffer is due to:

- The weak acid which provides H_3O^+ to trap added **OH⁻**.
- The salt of this acid which provides **A⁻** to trap added H_3O^+.

Action of a buffer solution

Go online

In an acid buffer, the weak acid supplies hydrogen ions when these are removed by the addition of a small amount of base. The salt of the weak acid provides the conjugate base, which can absorb hydrogen ions from addition of small amounts af acid.

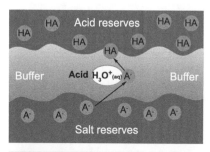

Addition of acid to the buffer

Extra hydrogen ions in the buffer upset the equilibrium situation in the weak acid.

$$HA \rightleftharpoons H_3O^+ + A^- \quad \text{equilibrum shifts} \leftarrow$$

The position of equilibrium shifts (Le Chatelier's principle) and the large reserves of **A⁻** ions from the salt allow the **H₃O⁺** ions to be removed. The **A⁻** ions provide the conjugate base.

Addition of hydroxide to the buffer

Extra hydroxide ions in the buffer react with some **H⁺** ions and upset the equilibrium situation in the weak acid.

$$\text{equilibrum shifts} \rightarrow \quad HA \rightleftharpoons H_3O^+ + A^-$$

TOPIC 1. CHEMICAL EQUILIBRIUM

> The position of equilibrium shifts (Le Chatelier's principle) and the large reserves of **HA** molecules from the weak acid allow the H_3O^+ ions to be restored.

1.5.2.2 Basic buffers

A basic buffer consists of a solution of a weak base and one of its salts, e.g. ammonia solution and ammonium chloride. The ammonia solution is partly ionised and the ammonium chloride is completely ionised. If hydrogen ions are added, they combine with ammonia and if hydroxide ions are added, they combine with the ammonium ions (conjugate acid) provided by the salt (NH_4Cl).

Figure 1.10: Equilibria in base buffer solutions

$$NH_3 + H_2O \rightleftharpoons NH_4^+ + OH^-$$

$$NH_4Cl \longrightarrow NH_4^+ + Cl^-$$

The stable pH of the buffer is due to:

- The weak base which provides NH_3 to trap added H_3O^+.
- The salt of this base which provides NH_4^+ to trap added OH^-.

Summary of buffer systems   Go online

Q67: Complete the formulae using the word bank to show how a basic buffer solution made from aqueous ammonia and ammonium chloride behave when acid or alkali are added.

Adding acid H_3O^+ + [_____] $\rightleftharpoons$ [_____]

Adding alkali OH^- + [_____] $\rightleftharpoons$ [_____] + [_____]

Word bank

| NH_3 | NH_4^+ | NH_4^+ | NH_3 | H_2O |

....

Q68: Complete the formulae using the word bank to show how an acidic buffer solution made from ethanoic acid and sodium ethanoate behaves when acid or alkali are added.

Adding acid H_3O^+ + [......] ⇌ [......]

Adding alkali OH^- + [......] ⇌ [......] + [......]

Word bank

| CH_3COOH | CH_3COO^- | CH_3COO^- | CH_3COOH | H_2O |

1.5.3 Calculating pH and buffer composition

A glance at Table 1.1 shows that biological buffer solutions have to operate around specific pH values. This pH value depends upon two factors. The acid dissociation constant and the relative proportions of salt and acid. The dissociation constant for a weak acid **HA** is given by this expression:

$$HA_{(aq)} \rightleftharpoons H_3O^+_{(aq)} + A^-_{(aq)}$$

$$K_a = \frac{[H_3O^+][A^-]}{[HA]}$$

$$[H_3O^+] = K_a \times \frac{[HA]}{[A^-]}$$

Two assumptions can be made that simplify this expression even further.

1. In a weak acid like **HA**, which is only very slightly dissociated, the concentration of **HA** at equilibrium is approximately the same as the molar concentration put into the solution.
2. The salt **MA** completely dissociates. Therefore **[A⁻]** will effectively be the concentration supplied by the salt.

The expression becomes:

$$[H_3O^+] = K_a \times \frac{[acid]}{[salt]}$$

taking the negative log of each side:

$$pH = pK_a - \log \frac{[acid]}{[salt]}$$

Two important points can be seen from this equation:

- Since $\frac{[acid]}{[salt]}$ is a ratio, adding water to a buffer will not affect the ratio (it will dilute each equally) and therefore will not affect the **[H_3O^+]** which determines the pH.

TOPIC 1. CHEMICAL EQUILIBRIUM

- If the **[acid] = [salt]** when the buffer is made up, the pH is the same as the pK_a (or $H_3O^+ = K_a$)

This equation allows calculation of pH of an acid buffer from its composition and acid dissociation constant, or calculation of composition from the other two values. Values for K_a and pK_a are available in the data booklet.

The next two problems show examples of the two most common type of calculation.

> **Example : Calculating pH from composition and K_a**
>
> Calculate the pH of a buffer solution made with 0.1 mol ℓ^{-1} ethanoic acid ($K_a = 1.7 \times 10^{-5}$) and sodium ethanoate if the salt is added:
>
> a) at 0.1 mol ℓ^{-1}
>
> b) at 0.2 mol ℓ^{-1}
>
> a) With the salt at 0.1 mol ℓ^{-1}
>
> $$[H_3O^+] = K_a \times \frac{[acid]}{[salt]}$$
> $$[H_3O^+] = 1.7 \times 10^{-5} \times \frac{0.1}{0.1}$$
> $$[H_3O^+] = 1.7 \times 10^{-5} \text{ mol}^{-1}$$
> $$pH = -\log[H_3O^+] = -\log(1.7 \times 10^{-5})$$
> $$pH = -(0.23 - 5.00)$$
> $$pH = 4.77$$
>
> b) With the salt at 0.2 mol ℓ^{-1}
>
> $$[H_3O^+] = K_a \times \frac{[acid]}{[salt]}$$
> $$[H_3O^+] = 1.7 \times 10^{-5} \times \frac{[0.1]}{[0.2]}$$
> $$[H_3O^+] = 1.7 \times 10^{-5} \times 0.5 \text{ mol l}^{-1}$$
> $$pH = -\log[H_3O^+] = -\log(8.5 \times 10^{-6})$$
> $$pH = -(0.93 - 6.00)$$
> $$pH = 5.07$$

Notice that doubling the salt concentration has only raised the pH by 0.3 (from 4.77 to 5.07) and in general, the pH of the buffer is tied closely to the pK_a value for the weak acid, in this case ethanoic acid $pK_a = 4.76$. The ratio of acid to salt effectively provides a 'fine tuning' of the pH.

© HERIOT-WATT UNIVERSITY

Example : Calculating composition from pH and pK$_a$

Calculate the concentrations of ethanoic acid and sodium ethanoate required to make a buffer solution with a pH of 5.3 (pK$_a$ in data booklet).

$$pH = pK_a - \log\frac{[acid]}{[salt]}$$

$$5.3 = 4.76 - \log\frac{[acid]}{[salt]}$$

$$\log\frac{[acid]}{[salt]} = -0.54$$

$$\log\frac{[salt]}{[acid]} = +0.54$$

$$\frac{[salt]}{[acid]} = 3.47$$

So the ratio of **3.47 to 1** is required and 3.47 moles of sodium ethanoate mixed with one litre of 1.0 mol ℓ^{-1} ethanoic acid could be used.

Buffer calculations   Go online

Q69: Calculate the pH of a buffer solution containing 0.10 mol ℓ^{-1} ethanoic acid and 0.50 mol ℓ^{-1} sodium ethanoate (K$_a$ is in the data booklet). Give your answer to two decimal places.

....

Q70: Calculate the composition of methanoic acid and sodium methanoate required to make a buffer solution with a pH of 4.0. Quote your answer as a ratio of salt to 1 (so 6.31 to 1 would quote as 6.31).

....

Q71: A 0.10 mol ℓ^{-1} solution of a weak acid has 0.40 mol ℓ^{-1} of its sodium salt dissolved in it. The resulting buffer has a pH 5.35. Find the dissociation constant of the acid.

....

Q72: What pH (to one decimal place) would be expected if 7.20g of sodium benzoate was dissolved in one litre of 0.02 mol ℓ^{-1} benzoic acid (sodium benzoate is C_6H_5COONa, K$_a$ and pK$_a$ in data booklet).

....

Q73: To prepare 1 litre of a buffer solution which would maintain a pH 5.5, 0.6g of ethanoic acid was used. What mass in grams of sodium ethanoate should the solution contain? Answer to two decimal places.

Q74: One of the systems which maintains the pH of blood at 7.40 involves the acid $H_2PO_4^-$ and the salt containing HPO_4^{2-}. Calculate the ratio of salt to acid in blood (K$_a$ and pK$_a$ of $H_2PO_4^-$ in data booklet). Assume that the weak acid is $H_2PO_4^-$ and the salt contains HPO_4^{2-}.

Extra questions

Q75: Ethanoic acid CH_3COOH is a weak acid. What is the conjugate base of ethanoic acid?

a) CH_3
b) OH^-
c) CH_3COO^-
d) H_3O^+

Q76: What is the conjugate base of H_2SO_4?

a) H_3O^+
b) OH^-
c) SO_4^{2-}
d) HSO_4^-

Q77: Lithium hydroxide is a strong base. What is the conjugate acid of lithium hydroxide?

a) Li^+
b) OH^-
c) H_3O^+
d) H_2O

Q78: Is a H_3O^+ concentration of 10^{-2} mol ℓ^{-1} greater or smaller than 10^{-12} mol ℓ^{-1}?

a) Greater
b) Smaller

Q79: As the concentration of the H_3O^+ ion increases the value of the pH:

a) Increase.
b) Decreases.

Calculate the pH of the following solutions to two decimal places.

Q80: A sample of water with $[H_3O^+] = 1.0 \times 10^{-7}$ mol ℓ^{-1}

Q81: A soft drink with $[H_3O^+] = 3.1 \times 10^{-4}$ mol ℓ^{-1}.

Q82: A blood sample with $[H_3O^+] = 4.0 \times 10^{-8}$ mol ℓ^{-1}

> **Exam hint:**
> Be careful to press the **log** or **log₁₀** key on your calculator to find the pH. Do not use the **ln** or **logₑ** key.

> **Key point**
>
> The pH scale shows whether a solution is acidic or alkaline.
>
> Solutions for which pH < 7 are acidic; pH 7 is neutral; and pH > 7 are alkaline.

The dissociation constant of water, K_w

Pure water dissociates as shown to a small extent:

$$H_2O(l) + H_2O(l) \rightleftharpoons H_3O^+(aq) + OH^-(aq)$$

Q83: Equilibrium constant is given by which expression?

a) $K_c = H_3O^+] [OH^-] / [H_2O]$
b) $K_c = [H_2O] [H_3O^+] / [OH^-]$
c) $K_c = [OH^-] [H_2O] / [H_3O^+]$

In this expression, $[H_2O]$ is effectively constant.

Q84: The ionic product of water with the special symbol K_w is given by which expression?

a) $K_w = [H_2O]$
b) $K_w = [H_2O] / [H_3O^+] [OH^-]$
c) $K_w = [H_3O^+] [OH^-]$
d) $K_w = [H_3O^+] [OH^-] / [H_2O]$

Q85: Using the word bank, complete the following sentences.

In _____ the concentrations of the ionic species $[H_3O^+]$ is equal to $[OH^-]$ and the value of K_w is 1.0×10 _____ at 25°C.

In acidic solutions the concentrations of the ionic species $[H_3O^+]$ is _____ $[OH^-]$ and the value of K_w is 1.0×10 _____ at 25°C.

In alkaline solution the concentrations of the ionic species $[H_3O^+]$ is _____ $[OH^-]$ and the value of K_w is 1.0×10 _____ at 25°C.

Word Bank: neutral water; greater than; less than; -2; -7; -14.

The acid dissociation constant, K_a and pK_a

Some acids in water dissociate only partially to form H_3O^+ ion and the conjugate base. These acids are called weak acids.

For a weak acid HA the equilibrium in solution is described by:

$$HA(aq) \rightleftharpoons H_3O^+(aq) + A^-(aq)$$

where A^- is the conjugate base.

The equilibrium constant K_a - called the acid dissociation constant - is

$$K_a = \frac{[H_3O^+][A^-]}{[HA]}$$

As with $[H_3O^+]$ and pH, the term pK_a has been invented to turn small numbers into more manageable ones.

$$pK_a = -\log_{10} K_a$$

Q86: The acid dissociation constant K_a of nitrous acid is 4.7×10^{-4}. What is the value of pK_a? Answer to one decimal place.

....

Q87: An acid has a pK_a of 9.8. What is the value of K_a?

....

Q88: The table lists the K_a values for four acids. Write the letters of these acids in order of decreasing acid strength (strongest first).

Letter	Name	K_a
A	Alanine	9.0×10^{-10}
B	Butanoic acid	1.5×10^{-5}
C	Carbonic acid	4.5×10^{-7}
D	Dichloroethanoic acid	5.0×10^{-2}

....

Q89: The table lists the pK_a values for four acids. Write the letters of these acids in order of decreasing acid strength (strongest first).

Letter	Name	pK_a
A	Benzoic acid	4.2
B	Boric acid	9.1
C	Bromoethanoic acid	2.9
D	Butanoic acid	4.8

Buffer solutions

A buffer solution is one that resists a change in pH when moderate amounts of an acid or base are added to it.

Let us imagine we add small amounts of acid to water. The pH of the water to begin with is 7 as it is neutral. The table shows the pH of a series of solutions prepared from adding successive 2 cm³ portions of 0.1 mol ℓ^{-1} acid to 100 cm³ of water.

Volume of acid added/cm³	0	2	4	6	8	10
pH	7.00	2.71	2.42	2.25	2.13	2.04

The pH drops a lot and the solution gets more acidic. Note particularly the drop in pH from 7.00 to 2.71 on adding the first 2 cm³ of acid.

Now consider the next table where the same volumes of acid are added to 100 cm³ of a buffer solution. The buffer solution was made up to have an initial pH of 5.07 using a mixture of ethanoic acid and sodium ethanoate in water.

Volume of acid added/cm³	0	2	4	6	8	10
pH	5.07	5.06	5.05	5.03	5.02	5.01

The data show that the pH of the buffer solution changes only slightly with the addition of the acid. Compared with water, it drops by only 0.01 pH units after the first 2 cm³ of acid is added.

Buffers are very useful because of this property. For many industrial processes and for biological situations it is important to keep a constant pH. For example, buffers exist in human blood, otherwise a change in pH of 0.5 unit might cause you to become unconscious.

TOPIC 1. CHEMICAL EQUILIBRIUM

The same number of H_3O^+ ions were added to both the water and the buffer solution. What happened to most of the H^+ ions added to the buffer? How does a buffer work? The secret lies in the mixture of ingredients. The acid buffer described is a solution of sodium ethanoate ($CH_3COO^- Na^+$) and ethanoic acid (CH_3COOH) in water.

Q90: With this combination is the pH acidic or basic? Type 'acidic' or 'basic'.

....

Q91: Will the pH be less than or greater than 7? Answer 'less' or 'greater'.

Ethanoic acid is a weak acid and sodium ethanoate is the sodium salt of ethanoic acid. Sometimes this is called the conjugate base. But more about that later.

A basic buffer is a solution of a weak base such as ammonia and the corresponding salt such as ammonium chloride.

Q92: Will the pH be less than or greater than 7? Answer 'less' or 'greater'.

> **Key point**
>
> Depending on the composition of the buffer solution, the solution can be acidic (pH < 7), neutral (pH = 7) or basic (pH > 7).

Returning to an acidic buffer such as sodium ethanoate/ethanoic acid solution, remember that sodium ethanoate is a salt and dissociates completely in aqueous solution;

$$CH_3COO^-Na^+(aq) \rightarrow CH_3COO^-(aq) + Na^+(aq)$$

to produce ethanoate ions.

Ethanoic acid, because it is a weak acid, dissociates only slightly;

$$CH_3COOH(aq) \rightleftharpoons CH_3COO^-(aq) + H_3O^+(aq)$$

and reaches an equilibrium where most of the ethanoic acid exists as undissociated molecules. Because, in the buffer solution, ethanoate ions are provided by sodium ethanoate, the equilibrium position shifts to form even more ethanoic acid molecules.

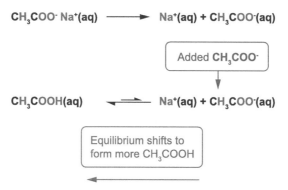

The result is that the buffer solution contains a reservoir of ethanoic acid molecules and ethanoate ions. As we shall see, this gives it the capability of reacting with either acid or base added to the buffer solution thus minimising their effect.

Addition of acid (H_3O^+)

When H_3O^+ is added to the buffer solution, the H_3O^+ reacts with the CH_3COO^- ions (present mainly from the sodium ethanoate) to form ethanoic acid. On account of this reaction, the H_3O^+ ion concentration remains virtually unchanged; the pH hardly alters.

$$CH_3COO^-(aq) + H_3O^+(aq) \rightarrow CH_3COOH(aq)$$

Addition of alkali (OH^-)

When OH^- ion is added to the buffer solution, the OH^- reacts with CH_3COOH (base plus acid forms a salt plus water);

$$CH_3COOH(aq) + OH^-(aq) \rightarrow CH_3COO^-(aq) + H_2O(l)$$

to form CH_3COO^- and, because the extra OH^- is removed, the pH of the buffer solution hardly changes.

TOPIC 1. CHEMICAL EQUILIBRIUM

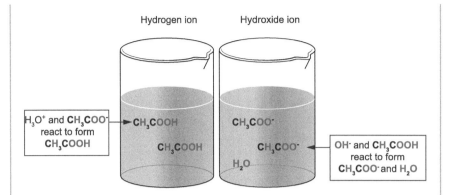

Conjugate acids and bases

There are a large number of buffer solutions. One was given as an example in this section. In general, one component of an acid buffer is a weak acid represented as **HA**. It will be only slightly dissociated. Large reserves of **HA molecules** are present in the buffer.

$$HA \rightleftharpoons H_3O^{\oplus} + A^{\ominus} \quad \text{at equilibrium}$$

The other component is a salt **MA** of the weak acid. This will be completely dissociated. Large reserves of the **A⁻ ion** are present in the buffer. This is the **conjugate base**.

$$MA \rightarrow M^{\oplus} + A^{\ominus} \quad \text{complete dissociation}$$

Consider an acidic buffer consisting of a solution of phosphoric acid (H_3PO_4) and potassium dihydrogenphosphate ($K^+H_2PO_4^-$)

Q93: When H_3O^+ is added to this buffer, which component will react with it?

a) H_3PO_4
b) $H_2PO_4^-$

....

Q94: What will form from this reaction?

Now let us consider a buffer containing a weak base and a salt.

The figure shows the two reactions relevant to a buffer solution containing ammonia and ammonium chloride, where ammonia is the weak base.

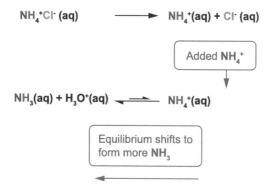

Q95: Is this buffer acidic or basic?

a) Acidic
b) Basic

....

Q96: Name the base in a buffer solution containing ammonia and ammonium chloride.

....

Q97: Name the conjugate acid in a buffer solution containing ammonia and ammonium chloride.

....

Q98: When acid is added to this buffer solution the H_3O^+ combine with:

a) Ammonia, NH_3
b) Ammonium ion, NH_4^+
c) Ammonium chloride, NH_4Cl

....

Q99: When alkali is added to this buffer solution the OH^- combine with:

a) Ammonia, NH_3
b) Ammonium ion, NH_4^+
c) Ammonium chloride, NH_4Cl

Q100: Using the word bank complete the paragraph.

An _____ buffer solution contains a mixture of a _____ and one of its _____. An example is a mixture of _____ and potassium ethanoate in water.

A _____ buffer solution contains a mixture of a _____ and one of its _____. An example is a mixture of _____ and ammonium chloride in water.

Word bank: salts; acidic; weak base; weak acid; ammonia; ethanoic acid; basic.

TOPIC 1. CHEMICAL EQUILIBRIUM

> **Key point**
>
> An acid buffer contains a weak acid and a salt of the same weak acid.
>
> A basic buffer contains a weak base and a salt of the same weak base.

Calculating the pH of buffer solutions

In order to calculate the pH of an acidic buffer solution we need to know:

- the acid dissociation constant K_a
- the relative concentrations of the acid and its salt

For an equilibrium reaction of ethanoic acid:

$$CH_3COOH(aq) \rightleftharpoons CH_3COO^-(aq) + H_3O^+(aq)$$

the equilibrium constant expression K_a is:

$$K_a = \frac{[H_3O^+][CH_3COO^-]}{[CH_3COOH]}$$

Note that the square brackets represent concentrations in moles per litre.

Calculate the molecular mass of ethanoic acid.

Q101: Relative formula mass of ethanoic acid. Enter the correct values from the SQA booklet into the boxes.

Formula CH_3COOH

which is $C_2H_4O_2$

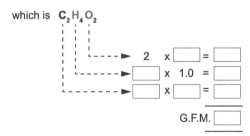

This number of grams when made up to 1 litre of aqueous solution will have a concentration of 1 moles per litre.

© HERIOT-WATT UNIVERSITY

Q102: How many grams of ethanoic acid would be required to dissolve in 1 litre of solution to make a 0.1 mol ℓ^{-1} solution?
i.e. [CH_3COOH] = 0.1
Answer to 1 decimal place without units.

....

Q103: At equilibrium at 25°C, the concentrations of each species in the 0.1 mol ℓ^{-1} solution are:

- [CH_3COO^-] = 1.304 × 10^{-3}
- [H^+] = 1.304 × 10^{-3}
- [CH_3COOH] = 0.100 - 0.0013 = 0.0987

Note that whilst we dissolved 0.1 moles of CH_3COOH in water only a small fraction of the molecules have dissociated to form CH_3COO^- and H_3O^+.

Calculate K_a for ethanoic acid at 25°C. Answer in standard form to 3 significant figures.

The following formula shows you the derivation of an equation to calculate the pH of a buffer solution. You will use this equation in the remaining questions in this section.

For the dissociation of an acid **HA**

$$HA \rightleftharpoons H_3O^+ + A^-$$

The equilibrium constant K_a is

$$K_a = \frac{[H_3O^+][A^-]}{[HA]}$$

Rearranging to make [H_3O^+] the subject

$$[H_3O^+] = K_a \frac{[HA]}{[A^-]}$$

Taking negative log_{10}

$$-log_{10}[H_3O^+] = -log_{10}K_a - log_{10}\frac{[HA]}{[A^-]}$$

Substituting $-log_{10}[H_3O^+]$ = **pH** and $-log_{10}K_a$ = pK_a

$$pH = pK_a - log_{10}\frac{[HA]}{[A^-]}$$

TOPIC 1. CHEMICAL EQUILIBRIUM

Mathematical relationships

Table 1.2: pH of buffers

[acid]/[salt]	10:1	2:1	**1:1**	1:2 (0.5:1)	1:10 (0.1:1)
$\log_{10}\frac{[acid]}{[salt]}$	1.0	0.30	0	-0.30	-1.0
pH	pK_a - 1.0	pK_a - 0.30	**pK_a**	pK_a + 0.30	pK_a + 1.0

The table shows that fine tuning of the pH of a buffer solution can be achieved by altering the ratio of [acid]/[salt]. For example, the pK_a of propanoic acid is 4.87 and thus the pH of a propanoic acid/sodium propanoate buffer would be 4.87 for a ratio of [acid]/[salt] = 1:1. (pH = pK_a)

Q104: A buffer solution is prepared using 0.1 mol ℓ^{-1} ethanoic acid and 0.2 mol ℓ^{-1} sodium ethanoate. Calculate the pH to two decimal places.

....

Q105: A buffer solution is prepared using 0.2 mol ℓ^{-1} ethanoic acid and 0.1 mol ℓ^{-1} sodium ethanoate. Calculate the pH to two decimal places.

....

Q106: A buffer consists of a aqueous solution of boric acid (H_3BO_3) and sodium borate (NaH_2BO_3). Use your data booklet to find the pK_a of boric acid. If the concentration of the acid was 0.2 mol ℓ^{-1} and the pH of the buffer solution was 9.4, what was the concentration of the sodium borate?

Try these two questions to ensure that you understand buffers.

Q107: You add 1.0 cm^3 of 0.2 mol ℓ^{-1} HCl to each of the following solutions. Which one will show the least change of pH?

a) 100 cm^3 of 0.1 mol ℓ^{-1} HCOOH
b) 100 cm^3 of 0.1 mol ℓ^{-1} HCOOH/0.2 mol ℓ^{-1} HCOO$^-$Na$^+$
c) 100 cm^3 of water

....

Q108: A buffer solution is prepared using 0.1 mol ℓ^{-1} ethanoic acid and 0.2 mol ℓ^{-1} sodium ethanoate. 1 cm^3 of the buffer solution is diluted with water to a final volume of 10 cm^3. Afterwards, does the pH increase, decrease or stay the same?

a) Increase
b) Decrease
c) Stay the same

1.6 Summary

Summary

You should now be able to:

- describe the equilibrium chemistry of acids and bases, write equilibrium expressions;
- describe the chemistry of acids and bases;
- use the terms: pH, K_w, K_a and pK_a;
- understand the chemistry of buffer solutions;
- calculate the pH of buffer solutions.

1.7 Resources

- pH Rainbow Tube (https://youtu.be/ZT9le3AUI7E)
- ChemConnections equilibrium (http://www.chemguide.co.uk)
- Davidson virtual chemistry experiments (http://bit.ly/29PGWyy)
- Bronsted Lowry Theory Of Acids And Bases (https://youtu.be/fM52LrQmel0)
- Bronsted Lowry Theory of Acids and Bases (https://youtu.be/0O2GeklXloA)

1.8 End of topic test

End of Topic 1 test   Go online

Q109: The value of K_c for a reaction at 500 K is 0.02. At 1000 K, it is 0.10.
Which of the following statements is true?

a) The reaction is endothermic.
b) The reaction is exothermic.
c) $\Delta H = 0$
d) The reaction is 5 times as fast at 1000 K as at 500 K.

....

Q110: Which of the following could have an equilibrium constant equal to 1×10^{-55}?

a) HCl (aq) + NaOH (aq) $\rightleftharpoons$ NaCl (aq) + H_2O (l)
b) Cu (s) + Mg^{2+} (aq) $\rightleftharpoons$ Cu^{2+} (aq) + Mg (s)
c) Zn (s) + $2Ag^+$ (aq) $\rightleftharpoons$ Zn^{2+} (aq) + 2Ag (s)
d) CH_3OH (l) + CH_3COOH (l) $\rightleftharpoons$ CH_3COOCH_3 (l) + H_2O (l)

TOPIC 1. CHEMICAL EQUILIBRIUM

Q111: Which of the following will increase the equilibrium constant for the following reaction given that ΔH, left to right, is positive?

$$N_2O_4(g) \rightleftharpoons 2NO_2(g)$$

a) Increase of pressure
b) Decrease of temperature
c) Use of a catalyst
d) Increase of temperature

Q112: In the reaction:

$$3O_2(g) \rightleftharpoons 2O_3(g)$$

K_p is 1×10^{-4}

What is the partial pressure of ozone, (O_3), if that of oxygen is 4 atm?

a) 2.67 atm
b) 8×10^{-2} atm
c) 4×10^{-4} atm
d) 1×10^{-4} atm

The Haber process is represented by the equation:

$$N_2\,(g) + 3H_2\,(g) \rightleftharpoons 2NH_3\,(g)$$

for which $\Delta H° = -92$ kJ mol^{-1}

2.0 moles of each reactant were allowed to react and come to equilibrium in a 1 litre container at 400 K. At equilibrium, 0.4 moles of ammonia were formed.

Q113: Which equilibrium constant expression is correct for this reaction?

a) $\dfrac{[H_2][N_2]}{[NH_3]}$
b) $\dfrac{[NH_3]}{[H_2][N_2]}$
c) $\dfrac{[NH_3]^2}{[H_2]^3[N_2]}$
d) $\dfrac{[H_2]^3[N_2]}{[NH_3]^2}$

Q114: Calculate the equilibrium concentration of nitrogen.

Q115: Calculate the equilibrium concentration of hydrogen.

Q116: Calculate the value of the equilibrium constant at this temperature.

Q117: Explain what will happen to the value of K, if the temperature is now raised to 600 K.

....

Q118: When a solute is shaken into two immiscible liquids, it partitions itself between the two liquids in a definite ratio. The value of this ratio is NOT dependent on the:

a) solute type.
b) temperature.
c) immiscible liquids involved.
d) mass of solute involved.

....

Q119: When a solute is distributed between two immiscible liquids and a partition coefficient of 2.0 is reached, the exchange rate between the two layers would be:

a) in a ratio of 2:1.
b) in a ratio of 1:2.
c) zero.
d) equal.

....

Q120: This diagram shows twelve moles of solute distributed between two immiscible liquids (one sphere represents one mole).

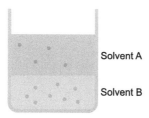

What would be the value of the partition coefficient solvent A / solvent B?

a) 0.5
b) 2
c) 4
d) 12

....

Q121: An aqueous solution of a monoprotic organic acid is shaken with ether at 25 °C until equilibrium is established.
20 cm^3 of the ether layer requires 15 cm^3 of 0.020 mol l^{-1} potassium hydroxide to neutralise and 20 cm^3 of the aqueous layer requires 7.5 cm^3 of 0.020 mol l^{-1} potassium hydroxide to neutralise.
Calculate the concentration of the organic acid in the ether layer.

....

Q122: Calculate the concentration of the organic acid in the aqueous layer.

....

Q123: Calculate the partition coefficient (ether/water) at 25°C.

....

Q124: An aqueous extract from a normal animal feedstuff and another from a feedstuff contaminated with a toxin were subjected to thin layer chromatography. The result is shown here.

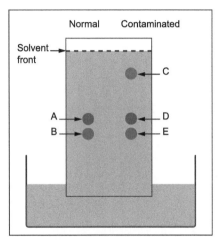

Which spot is most likely to be due to the toxin?

a) A
b) B
c) C
d) D
e) E

....

Q125: To prepare some toxin for further tests, a 50 cm^3 sample of the aqueous extract from the contaminated feedstuff was extracted with 50 cm^3 of hexane.

Using the information from the thin layer chromatography, which of the spots in the contaminated feedstuff is the least polar?

a) A
b) B
c) C
d) D
e) E

Q126: Into which layer would the toxin partition?

a) Water
b) Hexane

Q127: In which reaction is water behaving as a Bronsted-Lowry acid?

a) H_2O (l) + HF (aq) $\rightarrow$ H_3O^+ (aq) + F^- (aq)
b) H_2O (l) + NH_3 (aq) $\rightarrow$ NH_4^+ (aq) + OH^- (aq)
c) H_2O (l) + H_2SO_4 (l) $\rightarrow$ H_3O^+ (aq) + HSO_4^- (aq)
d) $2H_2O$ (l) + $2e^-$ $\rightarrow$ $2OH^-$ (aq) + H_2 (g)

Q128: For the reaction

$$NH_3 \text{ (aq)} + HCO_3^- \text{ (aq)} \rightarrow NH_4^+ \text{ (aq)} + CO_3^{2-} \text{ (aq)}$$

which of the following statements is correct?

a) NH_3 is the conjugate acid of NH_4^+.
b) HCO_3^- is the conjugate base of CO_3^{2-}.
c) CO_3^{2-} is the conjugate acid of HCO_3^-.
d) NH_4^+ is the conjugate acid of NH_3.

Q129: A solution of sulfuric acid is diluted until it has a concentration of 1×10^{-5} mol l^{-1}. Assuming that the acid is completely dissociated into ions, the pH would then be:

a) 1
b) 4.3
c) 4.7
d) 5

Q130: Sulfurous acid (H_2SO_3) is diprotic and dissociation occurs in two successive steps.

H_2SO_3 (aq) $\rightleftharpoons$ H_3O^+ (aq) + HSO_3^- (aq) $K_a = 1.5 \times 10^{-2}$
HSO_3^- (aq) $\rightleftharpoons$ H_3O^+ (aq) + SO_3^{2-} (aq) $K_a = 6.2 \times 10^{-8}$

Identify the **false** statements referring to the hydrogensulfite ion.

a) HSO_3^- is a stronger base than SO_3^{2-}.
b) HSO_3^- is the conjugate acid of SO_3^{2-}.
c) HSO_3^- is a weaker acid than H_2SO_3.
d) pK_a for HSO_3^- is less than pK_a for H_2SO_3.
e) HSO_3^- is amphoteric.

f) HSO_3^- is the conjugate base of H_2SO_3.

....

Q131: K_w is the ionic product of water.
Which of the following expressions correctly represents K_w?

a) $K_w = -\log([H_3O^+][A^-])$
b) $K_w = [H_3O^+][OH^-]$
c) $K_w = [H_3O^+][A^-]/[HA]$
d) $K_w = -\log[H_3O^+]$

....

Q132: The dissociation constant for water (K_w) varies with temperature.

$K_w = 0.64 \times 10^{-14}$ at 18°C
$K_w = 1.00 \times 10^{-14}$ at 25°C

From this information we can deduce that:

a) the ionisation of water is exothermic.
b) only at 25°C does the concentration of H_3O^+ equal the concentration of OH^-.
c) the pH of water is greater at 25°C than at 18°C.
d) water will have a greater electrical conductivity at 25°C than at 18°C.

....

Q133: 5.0 cm³ of a solution of hydrochloric acid was diluted to exactly 250 cm³ with water. The pH of this diluted solution was 2.0.
The concentration of the original undiluted solution was:

a) 0.02 mol l⁻¹
b) 0.50 mol l⁻¹
c) 0.40 mol l⁻¹
d) 0.04 mol l⁻¹

....

Q134: A solution with a pH of 3.2 has a **hydroxide** ion concentration which lies between:

a) 10^{-11} and 10^{-12} mol l⁻¹
b) 10^{-3} and 10^{-4} mol l⁻¹
c) 10^{-10} and 10^{-11} mol l⁻¹
d) 10^{-2} and 10^{-3} mol l⁻¹

....

Q135: Using information from the SQA data booklet, calculate the pH of a solution of ethanoic acid of concentration 0.01 mol l⁻¹.

....

Q136: A 0.05 mol l^{-1} solution of chloroethanoic acid has a pH of 2.19. Use this information to calculate pK$_a$ for chloroethanoic acid to 2 significant figures.

....

Q137: Indicators are weak acids for which the dissociation can be represented by:

$$HIn\ (aq) + H_2O\ (l) \rightleftharpoons H_3O^+\ (aq) + In^-\ (aq)$$
$\quad\quad$ Colour 1 $\quad\quad\quad\quad\quad\quad\quad\quad\quad\quad\quad\quad$ Colour 2

For any indicator, HIn, in aqueous solution, which of the following statements is correct?

a) The overall colour of the solution depends only on the pH.
b) In acidic solution, colour 2 will dominate.
c) The overall colour of the solution depends on the ratio of [HIn] to [In$^-$].
d) Adding alkali changes the value of K$_{in}$.

....

Q138: An indicator (H$_2$A) is a weak acid, and undergoes a two stage ionisation. The colours of the species are shown.

$$\underset{\text{Yellow}}{H_2A} \overset{\text{Stage 1}}{\rightleftharpoons} \underset{\text{Blue}}{HA^- + H_3O^+} \overset{\text{Stage 2}}{\rightleftharpoons} \underset{\text{Green}}{A^{2-} + 2H_3O^+}$$

The dissociation constants for the two ionisations are given by:

pK$_1$ = 3.5 and pK$_2$ = 5.9

Given that for an indicator pK = pH at the point where the colour change occurs, the indicator will be:

a) blue in a solution of pH 3 and green in a solution of pH 5.
b) yellow in a solution of pH 3 and blue in a solution of pH 5.
c) yellow in a solution of pH 3 and green in a solution of pH 5.
d) blue in a solution of pH 3 and blue in a solution of pH 5.

....

TOPIC 1. CHEMICAL EQUILIBRIUM

Q139: Which of the following graphs represents the change in pH as a strong alkali is added to a weak acid?

a)
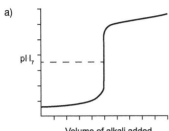
Volume of alkali added

b)
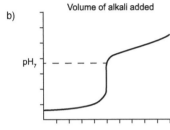
Volume of alkali added

c)
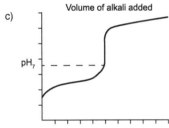
Volume of alkali added

d)
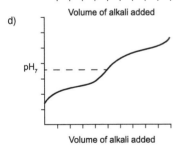
Volume of alkali added

....

Q140: A titration was carried out using potassium hydroxide solution and ethanoic acid. Which is the most suitable indicator for this titration?

a) Universal indicator, pH of colour change 4.0-11.0
b) Bromothymol blue indicator, pH of colour change 6.0-7.6
c) Phenolphthalein indicator, pH of colour change 8.0-9.8
d) Methyl orange indicator, pH of colour change 3.0-4.4

....

Q141: Why is it not practical to find the concentration of a solution of ammonia by titration with standard propanoic acid solution using an indicator?

a) The pH changes gradually around the equivalence point.
b) The salt of the acid and alkali does not have a pH of 7.
c) Organic acids are neutralised slowly by bases.
d) The salt of the acid and alkali is insoluble.

....

Q142: A buffer solution can be made from a:

a) strong acid and a salt of a weak acid.
b) weak acid and a salt of a strong acid.
c) weak acid and a salt of that acid.
d) strong acid and a salt of that acid.

....

Q143: 5 cm^3 of a 0.01 mol l^{-1} solution of hydrochloric acid was added to each of the following mixtures. The concentration of all the solutions is 0.1 mol l^{-1}.
In which case would there be the least change in pH?

a) 50 cm^3 NH_3 (aq) + 50 cm^3 HCl (aq)
b) 50 cm^3 NH_4Cl (aq) + 50 cm^3 NH_3 (aq)
c) 50 cm^3 HCl (aq) + 50 cm^3 Na Cl (aq)
d) 50 cm^3 NaCl (aq) + 50 cm^3 NH_4Cl (aq)

....

Q144: A buffer solution is made by dissolving 0.2 moles of sodium fluoride in one litre of hydrofluoric acid of concentration 0.1 mol l^{-1}.
Using information from the SQA data booklet, calculate the pH of the buffer solution.

....

Q145: A buffer solution of pH 4.6 was made up by dissolving sodium benzoate in a solution of benzoic acid.
If the concentration of the acid was 0.01 mol l^{-1}, calculate the concentration of the sodium benzoate, using information from the SQA data booklet.

TOPIC 1. CHEMICAL EQUILIBRIUM

When a pH electrode and meter are used to follow the titration between solutions of sodium hydroxide and methanoic acid, the following pH graph is obtained.

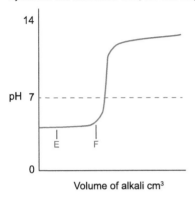

Volume of alkali cm³

Q146: Which of the following indicators could be used to detect the endpoint?

a) Methyl orange indicator, pH range 3.0 - 4.4
b) Methyl red indicator, pH range 4.2 - 6.3
c) Phenol red indicator, pH range 6.8 - 8.4
d) Alizarin yellow R indicator, pH range 10.0 - 12.0

....

Q147: If the titration is stopped between E and F, the resulting solution acts as a buffer. Why can this solution act as a buffer?

....

Q148: Explain how the pH of the buffer solution remains constant when a little more acid is added.

Unit 2 Topic 2

Reaction feasibility

Contents

- 2.1 Standard enthalpy of formation 155
- 2.2 Entropy 155
- 2.3 Second and third laws of thermodynamics 157
- 2.4 Free energy 159
- 2.5 Ellingham diagrams 162
    - 2.5.1 Extraction of metals 164
- 2.6 Summary 168
- 2.7 Resources 168
- 2.8 End of topic test 169

Prerequisites

Before you begin this topic, you should know:

- Standard Enthalpy of combustion - $\Delta H^{o}c$ is the energy released when one mole of a substance is burned completely in oxygen.

- Hess's Law - The energy change for a reaction is independent from the route taken to get from reactants to products.

- Bond Enthalpies - Energy required to break a bond and energy is given out when bonds are made.

- Thermochemistry is the study of energy changes that take place during chemical reactions.

Learning objective

By the end of this topic, you should understand:

- how Ellingham diagrams can be used to predict the conditions under which a reaction can occur;
- that Ellingham diagrams can be used to predict the conditions required to **extract** a metal from its oxide;
- standard conditions;
- standard enthalpy of formation - definition and relevant calculations $\Delta H° = \Sigma H°_f (p) - \Sigma H°_f (r)$;
- entropy and prediction of the effect on entropy of changing the temperature or state;
- the second and third law of thermodynamics;
- standard entropy changes. $\Delta S° = \Sigma S° (p) - \Sigma S° (r)$;
- the concept of free energy;
- the calculation of standard free energy change for a reaction $\Delta G° = \Sigma G° (p) - \Sigma G° (r)$;
- applications of the concept of free energy;
- the prediction of the feasibility of a reaction under standard and non-standard conditions ($\Delta G° = \Delta H° - T\Delta S°$);

TOPIC 2. REACTION FEASIBILITY

2.1 Standard enthalpy of formation

The **standard state** of a substance is the most stable state of the substance under **standard conditions** and the standard conditions refer to a pressure of one atmosphere and a specific temperature, usually 298 K (25°C).

> **Key point**
>
> In calculations it is important to use temperature in Kelvin (K). To convert from °C to K simply add 273.

The standard **enthalpy of formation** of a compound $\Delta H°_f$ is the energy given out or taken in when one mole of a compound is formed from its elements in their standard states.

For example the equation for the standard state of formation of ethanol is:

$$2C(s) + 3H_2(g) + \tfrac{1}{2} O_2(g) \rightarrow C_2H_5OH(l)$$

Standard enthalpy changes are measured in kJ mol^{-1} of reactant or product and the enthalpy change for a reaction is:

$$\Delta H° = \Sigma \Delta H°_{products} - \Sigma \Delta H°_{reactants}$$

where Σ = mathematical symbol sigma for summation.

Enthalpy values	ΔH sign	Reaction type
$\Sigma H(products) < \Sigma H(reactants)$	Negative	Exothermic
$\Sigma H(products) > \Sigma H(reactants)$	Positive	Endothermic

2.2 Entropy

The **entropy** of a system is the measure of the disorder within that system; the larger the entropy the larger the disorder and vice versa. Entropy is given the symbol **S** and the standard entropy of a substance S° is the entropy of one mole of the substance under standard conditions (1 atmosphere pressure and a temperature of 298K, 25°C). The units of entropy are J K^{-1} mol^{-1}.

Increasing entropy	Decreasing entropy
A puddle dries up on a warm day as the liquid becomes water vapour. The disorder (entropy) increases.	A builder uses a pile of loose bricks to construct a wall. The order of the system increases. Entropy falls.
Heating ammonium nitrate forms one mole of dinitrogen oxide and two moles of steam. Three moles of gas are formed. The disorder (entropy) increases.	When the individual ions in a crystal come together they take up a set position. The disorder of the system falls. Entropy decreases.
$NH_4NO_3(s) \rightarrow N_2O(g) + 2H_2O(g)$	$Na^+(aq) + Cl^-(aq) \rightarrow NaCl(s)$

© HERIOT-WATT UNIVERSITY

Entropy values of substances in the solid state tend to be low due to the particles in solid occupying fixed positions. The particles are unable to move, but vibrate. Gases have high entropy values since their particles have complete freedom to move anywhere within the space they occupy. Entropy values for liquids lies somewhere between that of solids and gases.

Entropy and temperature

At 0K the particles in a solid no longer vibrate and are perfectly ordered. This means the entropy of a substance at 0K is zero (third law of Thermodynamics see next section). As the temperature increases, the entropy of the solid increases until the melting point, where there is a rapid increase in entropy as the solid melts into a liquid. A greater increase in entropy is seen when the liquid boils to become a gas.

The following graph shows how the entropy of a substance varies with temperature.

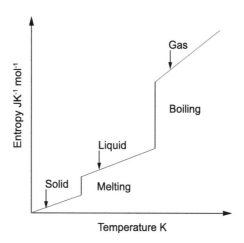

Calculating entropy in a chemical reaction

ΔS° = standard entropy change in a chemical reaction can be worked out using:-

$$\Delta S^\circ = \Sigma \Delta S^\circ_{(products)} - \Sigma \Delta S^\circ_{(reactants)}$$

Calculating entropy in a chemical reaction

Q1: Using the data book and the following information calculate the standard entropy change for the following reaction.

$$2AgNO_3(s) \rightarrow 2Ag(s) + 2NO_2(g) + O_2(g)$$

S° for $AgNO_3$ = 142 J K^{-1} mol^{-1}

S° for NO_2 = 241 J K^{-1} mol^{-1}

Remember to multiply the entropy value by the number of moles of the chemical involved.

2.3 Second and third laws of thermodynamics

One version of the second law of thermodynamics defines the conditions of a feasible reaction. It states that for a reaction to be feasible, the total entropy change for a reaction system and its surroundings must be positive (the total entropy must increase).

$$\Delta S°(\text{total}) = \Delta S°(\text{system}) + \Delta S° (\text{surroundings}) = +\text{ve}$$

Looking at the feasible reaction:

$$NH_3(g) + HCl(g) \rightarrow NH_4Cl(s)$$

$\Delta S°$(system) for this reaction is **-284** $J\ K^{-1}\ mol^{-1}$.

$\Delta H° = -176\ kJ\ mol^{-1}$ which means it is an exothermic reaction. The heat energy leaving the system causes the entropy of the surroundings to increase (hot surroundings have a higher entropy than cold surroundings).

$\Delta S°$(surroundings) = $-\Delta H°/T$ where T is the temperature taken as the standard temperature 298K.

So $\Delta S°$(surroundings) = -(-176)/298 = 0.591 $kJ\ K^{-1}\ mol^{-1}$ (591 $J\ K^{-1}\ mol^{-1}$)
Therefore $\Delta S°$(total) = $\Delta S°$(system) + $\Delta S°$ (surroundings)
= -284 + 591 = + 307 $J\ K^{-1}\ mol^{-1}$

The total entropy is positive confirming the reaction is feasible.

The second law of thermodynamics

Q2: Calcium carbonate, present in limestone, is stable under normal atmospheric conditions. When it is in a volcanic area and it gets very hot, it can thermally decompose. Given the following information, use the Second Law of Thermodynamics to show why limestone is stable at 25°C, but not at 1500°C.

$CaCO_3(s) \rightarrow CaO(s) + CO_2(g)$ $\Delta H° = +178\ kJ\ mol^{-1}$

$\Delta S° = +161\ J\ K^{-1}\ mol^{-1}$

....

Q3: Graphite has been converted into diamond by the use of extreme pressure and temperature. Given the following information and values of entropy in the data booklet, show why diamond can **not** be made from graphite at 1 atmosphere pressure, either at room temperature or 5000°C.

$C(s)_{graphite} \rightarrow C(s)_{diamond}$ $\Delta H = +2.0\ kJ\ mol^{-1}$

Since the entropy of a substance depends on the order of the system, when a solid crystal is cooled to absolute zero (zero Kelvin), all the vibrational motion of the particles is stopped with each particle having a fixed location, i.e. it is 100% ordered. The entropy is therefore zero. This is one version of the 'third law of thermodynamics'.

As temperature is increased, entropy increases. As with enthalpy values, it is normal to quote standard entropy values for substances as the entropy value for the standard state of the substance.

> **Key point**
>
> Notice that the unit of entropy values is joules per kelvin per mole ($J\ K^{-1}\ mol^{-1}$). Be careful with this, since enthalpy values are normally in kilojoules per mole ($kJ\ mol^{-1}$) and in problems involving these quantities the units must be the same.

An increase in entropy provides a driving force towards a reaction proceeding spontaneously. There are, however, processes that proceed spontaneously that seem to involve an entropy decrease.

For example, steam condenses to water at room temperature:

$$H_2O(g) \rightarrow H_2O(l) \qquad \Delta S° = 70 - 189 = -119\ J\ K^{-1}\ mol^{-1}$$
$$\Delta H° = -44.1\ kJ\ mol^{-1}$$

In this case, the enthalpy change is also negative. This outpouring of energy from the system ($-\Delta H_{SYSTEM}$) is transferred to the surroundings.

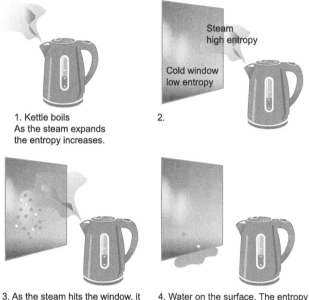

1. Kettle boils
As the steam expands the entropy increases.

2.

3. As the steam hits the window, it condenses. The window warms up and entropy increases.

4. Water on the surface. The entropy of the surroundings has increased.

The heat is transferred to the cold surface of the window and to the air around that area. This increases the disorder or entropy of the **surroundings**. (Just think of the scalding effect that would occur if your hand were placed in the steam - the disorder of the skin would increase!)

In general terms, heat energy released by a reaction system into the surroundings increases the entropy of the surroundings. If heat is absorbed by a reaction from the surroundings, this will decrease the entropy of the surroundings.

TOPIC 2. REACTION FEASIBILITY

In the case of steam condensing, the entropy gain of the surroundings is equal to the energy lost ($-\Delta H$) of the chemical system divided by the temperature:

$$\Delta S°_{SURROUNDINGS} = \frac{-\Delta H°_{SYSTEM}}{T}$$

The entropy change in the condensation situation therefore requires consideration of two entropy changes. The change in the system itself and the change in the surroundings must be added together, and for a spontaneous change to occur this total entropy change must be positive.

$$\Delta S°_{TOTAL} = \Delta S°_{SYSTEM} + \Delta S°_{SURROUNDINGS}$$

In the case of steam condensing, the heat given out can be used to calculate a value for $\Delta S°_{SURROUNDINGS}$:

$$\Delta S°_{SURR} = \frac{-\Delta H_{SYSTEM}}{T}$$
$$\Delta S°_{SURR} = \frac{-(-44.1 \times 1000)}{298}$$
$$\Delta S°_{SURR} = +148.0 \; JK^{-1} mol^{-1}$$

And a calculation of $\Delta S°_{SYSTEM}$ from the data booklet gives:

$$\Delta S°_{SYSTEM} = -119 \; JK^{-1} mol^{-1}$$
$$\text{therefore } \Delta S°_{TOTAL} = -119 + 148.0$$
$$\Delta S°_{TOTAL} = +29 \; JK^{-1} mol^{-1}$$

All this means that although the entropy of a system itself may drop, the process itself will still be a natural, spontaneous change if the drop is compensated by a larger increase in entropy of the surroundings.

Expressed another way, this is the **second law of thermodynamics**. The total entropy of a reaction system and its surroundings always increases for a spontaneous change.

A word of caution: spontaneous does not mean 'fast'. It means 'able to occur without needing work to bring it about'. Thermodynamics is concerned with the **direction** of change and not the **rate** of change.

2.4 Free energy

We know that:

$\Delta S°(total) = \Delta S°(system) + \Delta S°(surroundings)$

$\Delta S°(surroundings) = -\Delta H°/T$

From this $\Delta S°(total) = -\Delta H°/T + \Delta S°(system)$

Multiplying this expression by $-T$ we get $-T\Delta S°(total) = \Delta H° - T\Delta S°(system)$

$-T\Delta S°(total)$ has units of energy and we call this energy change the standard **free energy** change which is given the symbol $\Delta G°$.

$$\Delta G° = \Delta H° - T\Delta S°$$

© HERIOT-WATT UNIVERSITY

As $\Delta S°$(total) has to be positive for a reaction to be feasible $\Delta G°$ must be negative.

The equation can be used to predict whether a reaction is feasible or not.

$$2NaHCO_3(s) \rightarrow Na_2CO_3(s) + CO_2(g) + H_2O(g)$$

The decomposition of sodium hydrogen carbonate:

$\Delta H° = +129$ kJ mol^{-1}

$\Delta S° = +335$ J K^{-1} mol^{-1} (0.335 kJ K^{-1} mol^{-1})

At 298K (standard temperature) $\Delta G° = 129 - (298 \times 0.335) = +29$ kJ mol^{-1}

$\Delta G°$ at 298K is positive and therefore not feasible.

A reaction is feasible when $\Delta G°$ is negative and therefore becomes feasible when $\Delta G° = 0$

$$0 = \Delta H° - T\Delta S°$$

$$\text{Therefore } T = \Delta H°/\Delta S°$$

For the previous reaction the temperature at which it becomes feasible is 129/0.335 = 385K (112 °C).

The standard free energy change of a reaction can be calculated from the standard free energies of formation of the products and reactants.

$$\Delta G° = \Sigma \Delta G°_{f(products)} - \Sigma \Delta G°_{f(reactants)}$$

Calculations involving free energy changes

Q4: Use the table of standard free energies of formation to calculate values of $\Delta G°$ for these two reactions and thus predict whether or not the reaction is spontaneous.

a) $2Mg(s) + CO_2(g) \rightarrow 2MgO(s) + C(s)$
b) $2CuO(s) + C(s) \rightarrow 2Cu(s) + CO_2(g)$

SUBSTANCE	$\Delta G°_{FORMATION}$ / kJ mol^{-1}
CO_2	-394
MgO	-569
ZnO	-318
CuO	-130
All elements	0

Q5: Calculate the standard free energy change at both 400 K and 1000 K for the reaction:

	MgCO$_3$(s)	$\rightarrow$	MgO(s) +	CO$_2$(g)
$\Delta H°_f$ /kJ mol^{-1}	-1113		-602	-394
S°/J K^{-1} mol^{-1}	66		27	214

TOPIC 2. REACTION FEASIBILITY

Q6: Use the data given, along with data booklet values to calculate the temperature at which the Haber process becomes feasible.

	$N_2(g)$ +	$3H_2(g)$	$\rightleftharpoons$	$2NH_3(g)$
$\Delta H°_f$ /kJ mol^{-1}	0	0		-46.4
$S°$/J K^{-1} mol^{-1}	?	?		193.2

Q7: Given these reaction values for oxides of nitrogen:

$4NO(g) \rightarrow 2N_2O(g) + O_2(g)$ $\Delta G° = -139.56$ kJ

$2NO(g) + O_2(g) \rightarrow 2NO_2(g)$ $\Delta G° = -69.70$ kJ

a) Calculate $\Delta G°$ for this reaction:
$2N_2O(g) + 3O_2(g) \rightleftharpoons 4NO_2(g)$

b) Say whether the equilibrium position favours reactants or products.

Q8: Chloroform was one of the first anaesthetics used in surgery. At the boiling point of any liquid, the gas and liquid are in equilibrium. Use this information to calculate a boiling point for chloroform.

$CHCl_3(l) \rightarrow CHCl_3(g)$ $\Delta H° = 31.4$ kJ mol^{-1}

                                      $\Delta S° = 94.2$ J K^{-1} mol^{-1}

Free energy and equilibrium

$\Delta G°$ for a reaction can give information about the equilibrium position in a reversible reaction and the value of the equilibrium constant K.

$\Delta G° < 0$ the forward reaction will be feasible and therefore the products will be favoured over the reactants. The equilibrium position will lie to the right side (products) of the equilibrium and K will be greater than 1.

$\Delta G° > 0$ the backwards reaction will be feasible and therefore the reactants will be favoured over the products. The equilibrium position will lie to the left side (reactants) of the equilibrium and K will be less than 1.

Equilibrium reaction $R \rightleftharpoons P$

If $\Delta G°$ is negative and we start with 1 mole of pure R at 1 atmosphere of pressure standard state conditions apply and so at the start of the reaction we can talk about the standard free energy of R as opposed to the free energy of R.

As soon as the reaction starts and some R is converted to P standard state conditions no longer apply and therefore during a chemical reaction we talk about the free energy rather than the standard free energy. As we approach equilibrium the free energy moves towards a minimum. As $\Delta G°$ is negative the products are favoured and the equilibrium lies to the right (products).

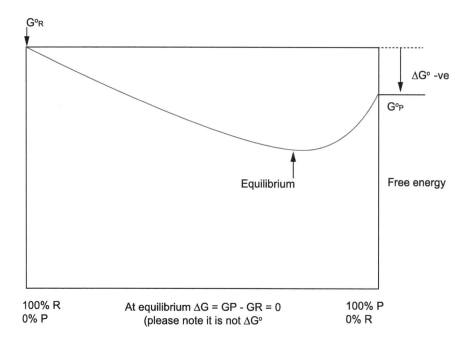

| 100% R | At equilibrium ΔG = GP - GR = 0 | 100% P |
| 0% P | (please note it is not ΔG°) | 0% R |

2.5 Ellingham diagrams

$\Delta G° = \Delta H° - T\Delta S°$ can be rearranged to $\Delta G° = -\Delta S°T + \Delta H°$.

Comparing this with the equation for a straight line y =mx+ c we can see from a plot of free energy change against temperature will have a gradient of $-\Delta S°$ and an intercept on the y-axis of $\Delta H°$. This is known as an Ellingham diagram.

Plotting an Ellingham diagram Go online

This table shows values of $\Delta G°$ over a temperature range for the formation of water gas (Equation 2.1).

Temperature K	200	400	600	800	1000	1200
$\Delta G°$ / kJ mol^{-1}	104	78	51	24	-3	-29

This graph shows values of $\Delta G°$ over a temperature range for the formation of water gas referring to:

$$C(s) + H_2O(g) \rightarrow CO(g) + H_2(g) \tag{2.1}$$

TOPIC 2. REACTION FEASIBILITY

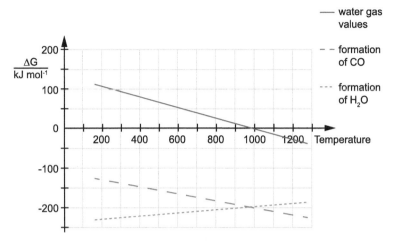

This reaction can be considered as a combination of these two oxide formations:

$$C(s) + {}^1/_2 O_2(g) \rightarrow CO(g) \qquad (2.2)$$

$$H_2(g) + {}^1/_2 O_2(g) \rightarrow H_2O(g) \qquad (2.3)$$

If the second equation (Equation 2.3) was reversed and added to the first (Equation 2.2) then the formation of water gas (Equation 2.1) results.

Both Equation 2.2 and Equation 2.3 have their values for $\Delta G°$ at various temperatures calculated and plotted onto the same graph.

Formation of CO(g)

Temperature K	200	400	600	800	1000	1200
$\Delta G°$ / kJ mol^{-1}	-128	-146	-164	-182	-200	-218

Formation of $H_2O(g)$

Temperature K	200	400	600	800	1000	1200
$\Delta G°$ / kJ mol^{-1}	-223	-224	-215	-206	-197	-188

Interpretation

The lines relating ΔG to temperature for the two equations Equation 2.2 and Equation 2.3 intersect at 981.3 K when $\Delta G = 0$. The carbon and hydrogen are both capable of reacting with oxygen but at any temperature **above** 981.3 K, the carbon is capable of winning oxygen from the water molecule and **forcing** the second equation (Equation 2.3) to reverse.

Consider 1000 K

$C(s) + \tfrac{1}{2}O_2(g) \rightarrow CO(g)$ $\Delta G° = -200$ kJ mol^{-1}
$H_2(g) + \tfrac{1}{2}O_2(g) \rightarrow H_2O(g)$ $\Delta G° = -197$ kJ mol^{-1}

By reversing the second equation and adding to the first, the result is:

$C(s) + H_2O(g) \rightarrow CO(g) + H_2(g)$ $\Delta G° = -3$ kJ mol^{-1}

At 1000 K the formation of water gas is feasible and spontaneous.

Consider 800 K

$C(s) + \tfrac{1}{2}O_2(g) \rightarrow CO(g)$ $\Delta G° = -182$ kJ mol^{-1}
$H_2(g) + \tfrac{1}{2}O_2(g) \rightarrow H_2O(g)$ $\Delta G° = -206$ kJ mol^{-1}

By reversing the second equation and adding to the first, the result is:

$C(s) + H_2O(g) \rightarrow CO(g) + H_2(g)$ $\Delta G° = +24$ kJ mol^{-1}

At 800 K the formation of water gas is **not** thermodynamically feasible or spontaneous. The Ellingham diagram provides a simple clear picture of the relationship between the different reactions and allows prediction of the conditions under which combinations of individual reactions become feasible.

2.5.1 Extraction of metals

Ellingham diagrams plot values of $\Delta G°$ against temperature. If the lines are drawn for metal oxide formation reactions, these can be used to predict the conditions required to **extract** a metal from its oxide. This requires the formation of the metal oxide process to be **reversed**. Any chemical used to aid the reversing of this process must provide enough free energy to supply this reversal. It is normal to write all reactions that are on the graph to involve one mole of oxygen (so that oxygen is removed when two equations are combined).

Interpreting Ellingham diagrams 1   Go online

Answer these questions on paper before displaying the explanation. In each case refer to the Ellingham diagram.

TOPIC 2. REACTION FEASIBILITY

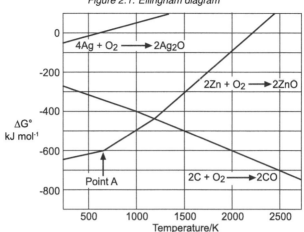

Figure 2.1: Ellingham diagram

Q9: Which oxide (in Figure 2.1) could be broken down by heat alone at 1000 K? (Hint: at 1000 K the ΔG value of the **reversed** reaction needs to be negative.)

....

Q10: Above which temperature would the breakdown of zinc oxide become feasible by heat alone?

....

Q11: Use the graph Figure 2.1 to calculate the $\Delta G°$ value for the reaction in which carbon reduces zinc oxide at:

a) 1000 K
b) 1500 K

$$2C(s) + 2ZnO(s) \rightarrow 2Zn(s) + 2CO(g)$$

Try each calculation for a) and b) on paper by following this route.

i. Write down the target equation.
ii. Write the equations for carbon combustion and zinc combustion, along with their $\Delta G°$ values from the graph at 1000 K in Figure 2.1.
iii. Write the reversed equation for the zinc combustion remembering to reverse $\Delta G°$.
iv. Add this new equation to the carbon equation and note the sign on $\Delta G°$. Is the reaction feasible or not?

....

Q12: At what temperature does the reduction of zinc oxide by carbon become feasible?

....

Q13: The data booklet gives the melting point of zinc as approximately 700 K. What happens to the entropy and what effect does it have on the gradient of the graph? (point A)
....

Q14: What causes the further little 'kink' in the zinc line at 1180 K?

Interpreting Ellingham diagrams 2

Q15: This Ellingham diagram shows the reactions involved in the blast furnace reduction of iron(II) oxide with carbon.

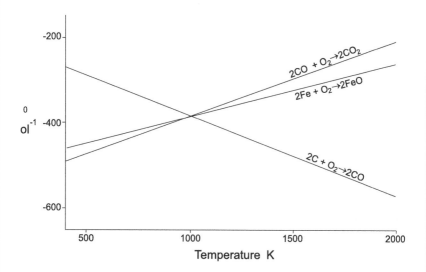

a) Write the combined equation showing the reduction of iron(II) oxide by carbon at 1500 K.
b) Calculate the standard free energy change at this temperature.
c) At what temperature does the reduction of iron(II) oxide by carbon become feasible?
d) At what temperatures will it be thermodynamically feasible for carbon monoxide to reduce iron(II) oxide?
e) Can you suggest a reason (apart from the temperature) why carbon monoxide might be more efficient than carbon at reducing iron(II) oxide?
....

Q16: Although magnesium ores are very abundant in the Earth's crust, the very high reactivity of magnesium makes it difficult to extract the metal. During the Second World War, magnesium was manufactured by reduction of its oxide by carbon.

$$2MgO + 2C \rightarrow 2CO + 2Mg$$

Examine the Ellingham diagram, Figure 2.2, and answer the questions which follow.

Figure 2.2: Ellingham diagram

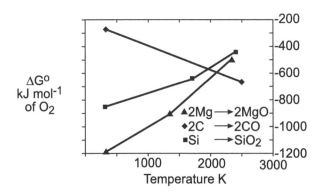

a) In what temperature range is the process thermodynamically feasible?
b) Describe **two** problems that the operation of the process at this temperature would present.
c) Use the Ellingham diagram to calculate $\Delta G°$ for the production of magnesium in the following:
$2MgO + Si \rightleftharpoons SiO_2 + 2Mg$ at 1500K
In industry, the extraction of magnesium from magnesium oxide using silicon involves two modifications.
  i. A mixture of calcium oxide and magnesium oxide is used, and the calcium oxide reacts with the silicon oxide produced.
$$CaO + SiO_2 \rightarrow CaSiO_3$$
($\Delta G° = -92$ kJ mol^{-1} at 1500 K)
  ii. The gaseous magnesium formed is continuously removed from the reaction mixture.
  iii. Use this information to answer the following questions.
d) Calculate $\Delta G°$ for the reaction:
$CaO + 2MgO + Si \rightarrow CaSiO_3 + 2Mg$ at 1500 K
e) Explain why the removal of magnesium from the reaction mixture helps the process.

2.6 Summary

Summary

You should now understand:

- how Ellingham diagrams can be used to predict the conditions under which a reaction can occur;
- that Ellingham diagrams can be used to predict the conditions required to **extract** a metal from its oxide;
- standard conditions;
- standard enthalpy of formation - definition and relevant calculations $\Delta H^\circ = \Sigma H^\circ_f \text{ (p)} - \Sigma H^\circ_f \text{ (r)}$;
- entropy and prediction of the effect on entropy of changing the temperature or state;
- the second and third law of thermodynamics;
- standard entropy changes. $\Delta S^\circ = \Sigma S^\circ \text{ (p)} - \Sigma S^\circ \text{ (r)}$;
- the concept of free energy;
- the calculation of standard free energy change for a reaction $\Delta G^\circ = \Sigma G^\circ \text{ (p)} - \Sigma G^\circ \text{ (r)}$;
- applications of the concept of free energy;
- the prediction of the feasibility of a reaction under standard and non-standard conditions ($\Delta G^\circ = \Delta H^\circ - T\Delta S^\circ$).

2.7 Resources

- Department of Physics University of Toronto: Einstein Quote (http://bit.ly/2aiFaui)
- BBC: The Second Law of Thermodynamics (http://bbc.in/29VpLMZ)
- ENDOTHERMIC reactions (http://bit.ly/29QsGos)
- ChemConnections (http://bit.ly/2awIv5g)
- Former AH Chemistry Unit 2 PPA 4, Verification of a Thermodynamic Prediction (http://bit.ly/2apLIFy)

2.8 End of topic test

End of Topic 2 test Go online

Q17: Which of the following reactions results in a large **decrease** in entropy?

a) $CaCO_3(s) \rightarrow CaO(s) + CO_2(g)$
b) $C(s) + H_2O(g) \rightarrow CO(g) + H_2(g)$
c) $N_2O_4(g) \rightarrow 2NO_2(g)$
d) $N_2(g) + 3H_2(g) \rightarrow 2NH_3(g)$

....

Q18: At 1400 K

$$2C + O_2 \rightarrow 2CO \quad \Delta G° = -475 \text{ kJ mol}^{-1} \text{ of } O_2$$
$$2Zn + O_2 \rightarrow 2ZnO \quad \Delta G° = -340 \text{ kJ mol}^{-1} \text{ of } O_2$$

For the reaction:

$$C + ZnO \rightarrow Zn + CO$$

the standard free energy change at 1400 K is:

a) 135 kJ mol^{-1} of ZnO
b) 67.5 kJ mol^{-1} of ZnO
c) -67.5 kJ mol^{-1} of ZnO
d) -135 kJ mol^{-1} of ZnO

....

Q19: The standard entropy values (J K^{-1} mol^{-1}) for a number of compounds are shown.

Compound	$CH_4(g)$	$O_2(g)$	$CO_2(g)$	$H_2O(g)$	$H_2O(l)$
S°	186	205	214	189	70

The standard entropy change (J K^{-1} mol^{-1}) for the complete combustion of one mole of methane is:

a) 242
b) -37
c) -242
d) -4

....

Q20: $\Delta G°$ gives an indication of the position of the equilibrium for a reaction.

The equilibrium lies on the side of the products when $\Delta G°$ is:

a) zero.
b) large and positive.
c) large and negative.
d) one.

....

Q21: At Tb, the boiling point of a liquid:

$$\Delta S_{vaporisation} = \frac{\Delta H_{vaporisation}}{T_b}$$

For many liquids:

$$\Delta S_{vaporisation} = 88 \text{ Jk}^{-1} \text{ mol}^{-1} \text{ (approx)}$$

If this value was true for water ($\Delta H°_{vaporisation}$ = 40.6 kJ mol^{-1}), the predicted boiling point of water would be:

a) 0.46 K
b) 373 K
c) 461 K
d) 2.16 K

Q22:

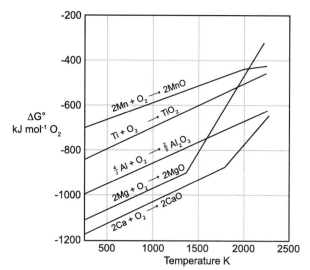

The reduction of Al$_2$O$_3$ to Al is thermodynamically feasible at 2000 K using:

a) titanium.
b) calcium.
c) magnesium.
d) manganese.

TOPIC 2. REACTION FEASIBILITY

Consider the following reactions and their values for $\Delta G°$ and $\Delta H°$ at 298 K.

Reaction	$\Delta G°$ (kJ mol^{-1})	$\Delta H°$ (kJ mol^{-1})
$1/2\ H_2(g) + 1/2\ Cl_2(g) \rightarrow HCl(g)$	-95	-92
$2\ Al(s) + 1\ 1/2\ O_2(g) \rightarrow Al_2O_3(s)$	-1576	-1669
$NH_4Cl(s) \rightarrow NH_4^+(aq) + Cl^-(aq)$	-7	16

Q23: The second reaction has the greatest difference in the values of $\Delta G°$ and $\Delta H°$. Suggest a reason for this difference.

....

Q24: From the values given for the third reaction, it can be concluded that ammonium chloride dissolves **spontaneously** in water under standard conditions with a drop in **temperature**.

Why can we come to these two conclusions?

....

Q25: Calculate the entropy change for the first reaction at 298 K.

....

Q26: Benzene boils at 80°C. The entropy change at this temperature is shown.

Calculate a value for the enthalpy change on boiling (the enthalpy of vaporisation).

$$C_6H_6(l) \rightleftharpoons C_6H_6(g)\ \Delta S°_{vaporisation} = 97.2\ J\ K^{-1}\ mol^{-1}$$

Give your answer in kJ mol^{-1} to one decimal place and with a sign.

Unit 2 Topic 3

Kinetics

Contents

3.1 Determination of order of reaction 174
3.2 Calculation of rate constants 175
3.3 Reaction mechanisms 177
3.4 Summary 181
3.5 Resources 181
3.6 End of topic test 182

Learning objective

By the end of this topic, you should know:

- how to determine the order of a reaction from experimental data and rate equations;
- how to calculate the rate constant and its units;
- how to predict from the rate equation the rate determining step and a possible mechanism.

3.1 Determination of order of reaction

Kinetics is about how fast a reaction goes (the rate of the chemical reaction).

The rate of a chemical reaction normally depends on the concentrations of the reactants.

$$A + B \rightarrow Products$$

If we double the initial concentration of A and keep the initial concentration of B constant the rate of reaction doubles. This suggests that the rate of reaction is directly proportional to the concentration of A so **rate** α **[A]^1**.

If the rate increases by a factor of four when the initial concentration of B is doubled and the initial concentration of A is kept constant it suggests that the rate is directly proportional to the square of the concentration of B so **rate** α **[B]^2**.

If we combine these results it gives us **rate** α **[A]^1[B]^2**.

This can be re-written as **rate = k [A]^1[B]^2** where k is the rate constant. This reaction would be **first** order with respect to A and **second** order with respect to B.

For a reaction of the type aA + bB $\rightarrow$ products we can express how the rate depends on the concentrations of A and B using the following expression **rate = k [A]^m[B]^n**. The indices m and n are the orders of the reaction with respect to A and B respectively (they bear no resemblance to the stoichiometric coefficients in balancing the chemical equation). They are usually small whole numbers no greater than 2. The overall order of the reaction is given by m + n so in the previous reaction the order would be 1 + 2 = 3, the reaction would be **third** order overall.

The rate constant k units depend on the overall order of the reaction.

Overall order	Units of k
0	mol l^{-1} s^{-1}
1	s^{-1}
2	mol^{-1} l s^{-1} (or l mol^{-1} s^{-1})
3	mol^{-2} l^2 s^{-1} (or l^2 mol^{-2} s^{-1})

Example:

rate = k [A]^2[B]^1

k = rate/[A]^2[B]^1

Rate measured in mol l^{-1} s^{-1} and concentration in mol l^{-1}

k = mol l^{-1} s^{-1} / (mol l^{-1})^2 (mol l^{-1})

k = mol l^{-1} s^{-1} / (mol^2 l^{-2}) (mol l^{-1})

k = mol^{-2} l^2 s^{-1}

3.2 Calculation of rate constants

The rate equation for a chemical reaction can only be determined experimentally. This is done through a series of experiments where the initial concentrations are varied. The initial rate for each experiment is determined.

$$A + B + C \rightarrow Products$$

Experiment	[A] mol l^{-1}	[B] mol l^{-1}	[C] mol l^{-1}	Initial rate mol l^{-1} s^{-1}
1	1.0	1.0	1.0	20
2	2.0	1.0	1.0	40
3	1.0	2.0	1.0	20
4	1.0	1.0	2.0	80

If we compare experiments 1 and 2 we can see that doubling the initial concentration of A causes the rate to increase by a factor of 2. This implies the reaction is first order with respect to A.

Comparing reactions 1 and 3 we can see that doubling the initial concentration of B has no effect on the initial rate of reaction implying that the reaction is zero order with respect to B.

Comparing reactions 1 and 4 we can see that doubling the initial concentration of C causes the initial rate of reaction to increase by a factor of 4. This implies the reaction is second order with respect to C.

The rate equation for this reaction would be rate = k $[A]^1[B]^0[C]^2$ more simply written as

Rate = k $[A]^1[C]^2$

The reaction is third order overall and the rate constant would have units of l^2 mol^{-2} s^{-1}.

To calculate the rate constant k we can use any one of the previous four reactions.

k = rate/$[A]^1[C]^2$ = 20/1.0 × (1.0)^2 = 20/1.0 = 20 l^2 mol^{-2} s^{-1}

Orders and rate constants Go online

The next two questions refer to the following reaction:

Bromide ions are oxidised by bromate ions (BrO$_3^-$) in acidic solution according to the equation:

$$5Br^-(aq) + BrO_3^-(aq) + 6H^+(aq) \rightarrow 3Br_2(aq) + 3H_2O(\ell)$$

By experiment, the reaction is found to be first order with respect to both bromide and bromate but second order with respect to hydrogen ions.

Q1: Write the rate equation for this reaction.

....

Q2: What is the overall order of the reaction?

The next four questions refer to the hydrolysis of urea in the presence of the enzyme, urease.

$$NH_2CONH_2(aq) + H_2O(l) \rightarrow 2NH_3(g) + CO_2(g)$$

The rate equation for the reaction is found by experiment to be:

$$\text{Rate} = k\,[\text{urea}][\text{urease}]$$

Q3: What is the overall order of reaction?

....

Q4: What is the order with respect to water?

....

Q5: What is the order with respect to urea?

....

Q6: What is the order with respect to urease?

The next four questions refer to the decomposition of dinitrogen pentoxide, N_2O_5:

$$2N_2O_5(g) \rightarrow 4NO_2(g) + O_2(g)$$

Experiments were carried out in which the initial concentration was changed and the initial rate of reaction was measured. The following data were obtained.

$[N_2O_5]$ / mol ℓ^{-1}	Initial Rate / mol ℓ^{-1} s^{-1}
0.05	2.2×10^{-5}
0.10	4.4×10^{-5}
0.20	8.8×10^{-5}

Q7: Write the rate equation for the reaction.

....

Q8: What will be the units of the rate constant?

....

Q9: Calculate the rate constant for the reaction. (Do not include units.)

....

Q10: If the initial concentration was 0.07 mol ℓ^{-1}, calculate the initial rate of reaction in mol ℓ^{-1} s^{-1}. (Do not include units.)

TOPIC 3. KINETICS

The next three questions refer to the following reaction:

Iodide ions are oxidised in acidic solution to triiodide ions, I_3^-, by hydrogen peroxide.

$$H_2O_2(aq) + 3I^-(aq) + 2H^+(aq) \rightarrow I_3^-(aq) + 2H_2O(\ell)$$

The following initial rate data were obtained:

Experiment	Initial Concentrations / mol ℓ^{-1}			Initial Rate / mol ℓ^{-1} s^{-1}
	[H_2O_2]	[I^-]	[H^+]	
1	0.02	0.02	0.001	9.2×10^{-6}
2	0.04	0.02	0.001	1.84×10^{-5}
3	0.02	0.04	0.001	1.84×10^{-5}
4	0.02	0.02	0.002	9.2×10^{-6}

Q11: From the data, write the rate equation.

....

Q12: What will be the units of k?

....

Q13: Calculate the value for the rate constant.

3.3 Reaction mechanisms

Chemical reactions usually happen by a series of steps rather than by one single step. This series of steps is known as the reaction mechanism. The overall rate of a reaction is dependent on the slowest step, which is called the **rate determining step**.

$$2NO_2 + F_2 \rightarrow 2NO_2F$$

$$\text{Rate} = k\,[NO_2][F_2]$$

The reaction is first order with respect to each of the reactants and this suggests that 1 molecule of each of the reactants must be involved in the slow rate determining step.

$$NO_2 + F_2 \rightarrow NO_2F + F \text{ (slow step)}$$

$$NO_2 + F \rightarrow NO_2F \text{ (fast step)}$$

Adding the two steps together gives the overall equation for the reaction.

Please note that an experimentally determined rate equation can provide evidence but not proof for a proposed reaction mechanism.

Do we need to know how fast each step is in order to work out the overall rate? As an analogy, consider this production line in the bottling plant in a distillery.

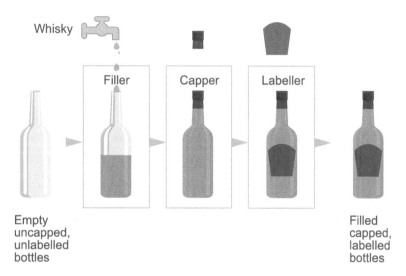

A whisky production line

There are three parts and the maximum capacity of each is:

Part 1	Filler	2 bottles filled per minute.
Part 2	Capper	120 bottles capped per minute.
Part 3	Labeller	60 bottles labelled per minute.

The production line is switched on. After 30 seconds a bottle has been filled and is passed to the Capper which caps it in 0.5 seconds and passes it to the Labeller which takes a further second to label it. So after 31.5 seconds we have completed one bottle. After one minute, the second bottle is full and is immediately capped by the Capper, which has been twiddling its thumbs waiting for the bottle to arrive.

Reaction mechanisms

Q14: How many seconds will it take to fill, cap and label 100 bottles?

....

Q15: How many seconds will it take to fill, cap and label 1000 bottles?

TOPIC 3. KINETICS

Clearly it does not matter how fast the Capper and Labeller are. The overall process is governed by how fast the bottles are filled. Bottle filling is the rate determining step.

In any chemical reaction mechanism, one step will be significantly slower than the others and this step will determine the overall reaction rate, i.e. it will be the rate determining step (RDS for short).

> **Key point**
>
> In general, the overall rate of a reaction depends on the rate of the slowest step in the mechanism. The rate equation provides information about the rate-determining step.

Questions on reaction mechanisms

Q16: Which of the following reactions is **most** likely to occur by a simple one-step process?

a) $4HBr + O_2 \rightarrow 2H_2O + 2Br_2$
b) $H_2S + Cl_2 \rightarrow S + 2HCl$
c) $2NO + O_2 \rightarrow 2NO_2$
d) $2H_2 + O_2 \rightarrow 2H_2O$

....

Q17: Which of the following reactions is **least** likely to occur by a simple one-step process?

a) $4HBr + O_2 \rightarrow 2H_2O + 2Br_2$
b) $H_2S + Cl_2 \rightarrow S + 2HCl$
c) $2NO + O_2 \rightarrow 2NO_2$
d) $2H_2 + O_2 \rightarrow 2H_2O$

The next six questions refer to the reaction between propanone and bromine in alkaline solution.

The balanced equation is:

$$CH_3COCH_3(aq) + Br_2(aq) + OH^-(aq) \rightarrow CH_3COCH_2Br(aq) + H_2O(\ell) + Br^-(aq)$$

The experimentally determined rate equation is:

$$\text{Rate} = k\,[CH_3COCH_3][OH^-]$$

Use this information to select True or False for each of the following statements.

Q18: The reaction is first order with respect to bromine.

a) True
b) False

....

Q19: The reaction involves a simple one-step process.

a) True
b) False

....

Q20: The reaction is second order overall.

a) True
b) False

Q21: The rate determining step involves one molecule of propanone and one molecule of bromine.

a) True
b) False

Q22: The following mechanism fits the rate equation.

$$H_3C-\underset{\underset{}{\overset{\overset{O}{\|}}{C}}}{}-CH_3 \longrightarrow H_3C-\underset{\underset{}{\overset{\overset{OH}{|}}{C}}}{}=CH_2 \qquad \textbf{slow}$$

$$H_3C-\underset{\underset{}{\overset{\overset{OH}{|}}{C}}}{}=CH_2 \longrightarrow H_3C-\underset{\underset{Br}{|}}{\overset{\overset{OH}{|}}{C}}-CH_2Br \qquad \textbf{fast}$$

$$H_3C-\underset{\underset{Br}{|}}{\overset{\overset{OH}{|}}{C}}-CH_2Br + {}^{\ominus}OH \longrightarrow H_3C-\underset{}{\overset{\overset{O}{\|}}{C}}-CH_2Br + H_2O + Br^{\ominus} \qquad \textbf{fast}$$

a) True
b) False

Q23: The following mechanism fits the rate equation.

$$H_3C-\underset{}{\overset{\overset{O}{\|}}{C}}-CH_3 + {}^{\ominus}OH \longrightarrow H_3C-\underset{}{\overset{\overset{O}{\|}}{C}}-CH_2^{\ominus} + H_2O \qquad \textbf{slow}$$

$$H_3C-\underset{}{\overset{\overset{O}{\|}}{C}}-CH_2^{\ominus} + Br_2 \longrightarrow H_3C-\underset{}{\overset{\overset{O}{\|}}{C}}-CH_2Br + Br^{\ominus} \qquad \textbf{fast}$$

a) True
b) False

TOPIC 3. KINETICS

The following questions refer to a reaction involving hydrogen peroxide and bromide ions in aqueous solution.

$$H_2O_2 + Br^- \rightarrow BrO^- + H_2O \quad \text{Step 1}$$
$$H_2O_2 + BrO^- \rightarrow Br^- + H_2O + O_2 \quad \text{Step 2}$$

Q24: What is the equation for the overall reaction?

....

Q25: What is the role played by the Br^- ion?

....

Q26: What role is played by the BrO^- ion?

....

Q27: If step 1 is the rate determining step, which of the following is the rate equation?

a) Rate = k $[H_2O_2]$
b) Rate = k $[H_2O_2][Br^-]$
c) Rate = k $[H_2O_2]^2$
d) Rate = k $[H_2O_2]^2 [Br^-]$

3.4 Summary

Summary

You should now be able to:

- determine the order of a reaction from experimental data and rate equations;
- calculate the rate constant and its units;
- predict from the rate equation the rate determining step and a possible mechanism.

3.5 Resources

- SSERC Bulletin (page 9) (http://bit.ly/29NFDnF)
- Finding orders of reaction (http://bit.ly/29QxVok)
- Royal Society of Chemistry: Chemistry at the Races (http://rsc.li/29WOApS)

3.6 End of topic test

End of Topic 3 test   Go online

Q28: Which of the following is the unit for the rate of a chemical reaction?

a) mol l^{-1}
b) s mol^{-1}
c) mol l^{-1} s^{-1}
d) s^{-1}

The following data refer to initial reaction rates obtained for the reaction.

$$X + Y + Z \rightarrow \text{products}$$

Experiment	Rate concentration			Relative initial rate
	[X]	[Y]	[Z]	
1	1.0	1.0	1.0	0.3
2	1.0	2.0	1.0	0.6
3	2.0	2.0	1.0	1.2
4	2.0	1.0	2.0	0.6

Q29: Which rate equation does this data fit?

a) Rate = k [X] [Y]
b) Rate = k [X] [Y] [Z]
c) Rate = k [X] [Y]^2
d) Rate = k [X]

....

Q30: The rate of a particular chemical reaction is first order with respect to each of two reactants. The units of k, the rate constant, for the reaction are:

a) l mol^{-1} s^{-1}
b) l^2 mol^{-2} s^{-2}
c) mol l^{-1} s^{-1}
d) mol^2 l^{-2} s^{-2}

....

Q31: 2X + 3Y$_2$ → products

A correct statement which can be made about the reaction is that:

a) the reaction will be slow due to the number of particles colliding.
b) the reaction order with respect to X is 2.
c) the overall order of the reaction is 5.
d) the rate expression cannot be predicted.

....

Q32: P + Q → R

The rate equation for this reaction is:

$$\text{Rate} = k[P][Q]^2$$

If the concentration of P and Q are both doubled, how many times will the rate increase?

a) 2
b) 4
c) 6
d) 8

....

Q33: The reaction expressed by the stoichiometric equation **Q + R → X + Z** was found to be first order with respect to each of the two reactants.

Which of the following statements is correct?

a) The rate of the reaction is independent of either Q or R.
b) The rate of reaction decreases as the reaction proceeds.
c) Overall, the reaction is first order.
d) If the initial concentrations of both Q and R are halved, the rate of reaction will be halved.

....

Q34: The reaction **X + 2Y → Z** has a rate equation of the form:

$$\text{Rate} = k[X][Y]$$

If the reaction proceeds by a two step process, then the rate-determining step might be:

a) **X + Y → Z**
b) **X + 2Y → intermediate**
c) **X + Y → intermediate**
d) **XY + Y → Z**

Q35: For the reaction **NO(g) + N$_2$O$_5$(g) → 3NO$_2$(g)** the following mechanism is suggested.

Step 1: N$_2$O$_5$(g) → NO$_2$ + NO$_3$(g) **slow**

Step 2: NO(g) + NO$_3$(g) → 2NO$_2$(g) **fast**

Experimental evidence to support this would be obtained if the rate of the reaction equals:

a) k[NO]
b) k[N$_2$O$_5$][NO]
c) k[NO][NO$_3$]
d) k[N$_2$O$_5$]

....

Q36: For a given chemical change involving two reactants P and Q, rate of reaction is directly proportional to [P][Q].

If the equation representing the overall reaction is **P + 2Q → S + T** the mechanism could be:

a) 2Q → R + S fast
   R + P → T slow
b) P + Q → R + S fast
   R + Q → T slow
c) P + Q → R + S slow
   R + Q → T fast
d) P → R + S fast
   R + 2Q → T slow

....

Q37: A reaction **A + B → C + D** is found to be first order with respect to B and zero order with respect to A.

Which of the following graphs is consistent with these results?

a)
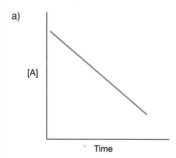

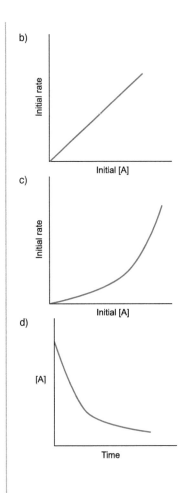

Mercury(II) chloride is reduced by oxalate ions ($C_2O_4^{2-}$) according to the equation:

$2HgCl_2 \text{ (aq)} + C_2O_4^{2-} \text{ (aq)} \rightarrow 2Cl^- \text{ (aq)} + 2CO_2 \text{ (g)} + 2HgCl \text{ (s)}$

The following data were obtained in a series of four experiments at the same temperature. The rate is measured by the decrease in concentration of $HgCl_2$ (aq) per minute.

Experiment	Initial concentration of $HgCl_2$(aq) / mol l^{-1}	Initial concentration of $C_2O_4^{2-}$ (aq) / mol l^{-1}	Initial rate / mol l^{-1} min^{-1}
1	0.128	0.304	1.82 x 10^{-4}
2	0.064	0.608	3.66 x 10^{-4}
3	0.128	0.608	7.31 x 10^{-4}
4	0.064	0.304	0.90 x 10^{-4}

Q38: From the data given, deduce the overall rate equation for the reaction.

....

Q39: Using the results for experiment 1 and your answer to part (a), calculate the rate constant at the given temperature.

....

Q40: What will be the units of k?

a) l mol^{-1} min^{-1}
b) mol min^{-1}
c) mol^2 l^{-2} min^{-1}
d) l^2 mol^{-2} min^{-1}
e) mol l^{-1} min^{-1}
f) l min^{-1}

....

Q41: Calculate the initial rate of the reaction when the initial concentration of each reactant is 0.1 mol l^{-1}.

Unit 2 Topic 4
Physical chemistry test

Physical chemistry test

Go online

Q1: $2SO_2 (g) + O_2 (g) \rightleftharpoons 2SO_3 (g)$

Removing the sulfur trioxide produced in the system will:

a) increase the value of the equilibrium constant.
b) decrease the concentration of SO_2 and O_2.
c) increase the concentration of O_2 only.
d) decrease the value of the equilibrium constant.
e) decrease the concentration of SO_2 only.
f) increase the concentration of SO_2 and O_2.

....

Q2: Which of the following would **not** act as a buffer solution?

a) Ethanoic acid and sodium ethanoate
b) Hydrochloric acid and sodium chloride
c) Sulfurous acid and potassium sulfite
d) Aqueous ammonia and ammonium chloride

500 cm^3 of a buffer solution contains 0.20 mol of ethanoic acid and 0.25 mol of sodium ethanoate.

Q3: Explain how this solution acts as a buffer on the addition of a small volume of potassium hydroxide solution.

....

Q4: Using the preceding data and information from the data booklet, calculate the pH of the buffer solution using the equation pH = pKa - log ([acid] / [salt]).

Magnesium nitrate decomposes on heating.

$$2Mg(NO_3)_2 (s) \rightarrow 2MgO (s) + 4NO_2 (g) + O_2 (g) \quad \Delta H° = + 510 \text{ kJ mol}^{-1}$$

Compound	S° J K^{-1} mol^{-1}
$Mg(NO_3)_2$ (s)	164
MgO (s)	27
NO_2 (g)	240
O_2 (g)	205

Q5: Calculate the standard entropy change for the reaction.

....

Q6: Calculate the temperature at which this reaction becomes feasible.

....

Q7: At which temperature would the entropy of a perfect crystal be zero?

a) 298 K
b) 273 K
c) 100 K
d) 0 K

....

Q8: $\Delta G°$ gives an indication of the position of the equilibrium for a reaction. The equilibrium lies on the side of the products when $\Delta G°$ is:

a) negative.
b) zero.
c) one.
d) positive.

....

Q9: The following reaction is first order with respect to each of the reactants.

$$A + B \rightarrow C + D$$

Which of the following is a correct statement about this reaction?

a) As the reaction proceeds its rate will increase.
b) The rate of reaction is independent of the concentration of either A or B.
c) The reaction is second order overall.
d) The reaction is first order overall.

The following kinetic data was obtained for the reaction:

$$P + 3Q \rightarrow R$$

[P] (mol l^{-1})	[Q] (mol l^{-1})	Initial rate of formation of R (mol l^{-1} min^{-1})
1.0	1.0	0.150
2.0	1.0	0.300
2.0	0.5	0.075

Q10: Deduce the rate equation for the reaction.

....

Q11: Predict an equation for the rate determining step.

....

Q12: Calculate the rate constant.

....

Q13: What will be the units of k?

a) l mol^{-1} min^{-1}
b) mol^2 l^{-2} min^{-1}
c) mol l^{-1} min^{-1}
d) l^2 mol^{-2} min^{-1}
e) l min^{-1}
f) mol min^{-1}

....

Q14: The reaction A + B → C has a rate law of the form rate = k[A][B].

If the concentration of A and B are both doubled, the rate will increase by a factor of:

a) 2
b) 4
c) 6
d) 8

Researching Chemistry

1	**Common chemical apparatus**	**193**
	1.1 Overview	194
2	**Skills involved in experimental work**	**195**
	2.1 Overview	196
3	**Stoichiometric calculations**	**197**
	3.1 Stoichiometry	198
	3.2 Stoichiometric calculation: preparation of benzoic acid by hydrolysis of ethylbenzoate	198
	3.3 Reactant in excess	200
	3.4 Stoichiometric calculation: nickel(II) ions and dimethylglyoxime	201
	3.5 Calculations associated with pharmaceuticals	202
	3.6 Summary	202
	3.7 End of topic test	203
4	**Gravimetric analysis**	**205**
	4.1 Gravimetric analysis	206
	4.2 Gravimetric analysis experiments	206
	4.3 Summary	209
5	**Volumetric analysis**	**211**
	5.1 Volumetric analysis	213
	5.2 Preparation of a standard solution	214
	5.3 Use of controls in chemical reactions	216
	5.4 Complexometric titrations	216
	5.5 Back titrations and associated calculations	218
	5.6 Summary	220
	5.7 Resources	220
	5.8 End of topic test	220
6	**Practical skills and techniques**	**223**
	6.1 Colorimetry and accurate dilution	224
	6.2 Distillation	225
	6.3 Refluxing	226
	6.4 Vacuum filtration	227

6.5	Recrystallisation	229
6.6	Use of a separating funnel	229
6.7	Thin-layer chromatography (TLC)	230
6.8	Melting point and mixed melting point	233
6.9	Summary	234
6.10	Resources	235
6.11	End of topic test	235

7 Researching chemistry test **239**

Unit 3 Topic 1

Common chemical apparatus

Contents
  1.1 Overview 194

> **Learning objective**
>
> Candidates must be familiar with the use(s) of the following types of apparatus:
>
> - conical flask;
> - digital balance;
> - pipette with safety filler;
> - burette;
> - volumetric (standard) flask;
> - distillation (round-bottomed) flask;
> - condenser;
> - thermometer;
> - Buchner or Hirsch or sintered glass funnel;
> - glassware with ground glass joints ('Quickfit' or similar);
> - thin-layer chromatography apparatus;
> - colorimeter;
> - melting point;
> - separating funnel.

1.1 Overview

The knowledge for this topic of study is embedded throughout the Researching Chemistry section of the course. The learning objectives are presented for reference.

Unit 3 Topic 2

Skills involved in experimental work

Contents
2.1 Overview 196

Learning objective

Candidates must be able to:

- tabulate data using appropriate headings and units of measurement;
- represent data as a scatter graph with suitable scales and labels;
- sketch a line of best fit (straight or curved) to represent the trend observed in the data;
- calculate average (mean) values;
- identify and eliminate rogue points;
- qualitatively appreciate the relative accuracy of apparatus used to measure the volume of liquids;
- comment on the reproducibility of results where measurements have been repeated;
- carry out quantitative stoichiometric calculations;
- interpret spectral data;
- appropriately use a positive control, for example a known substance, to validate a technique or procedure.

2.1 Overview

The knowledge for this topic of study is embedded throughout the Researching Chemistry section of the course. The learning objectives are presented for reference.

Unit 3 Topic 3

Stoichiometric calculations

Contents

3.1 Stoichiometry 198
3.2 Stoichiometric calculation: preparation of benzoic acid by hydrolysis of ethylbenzoate 198
3.3 Reactant in excess 200
3.4 Stoichiometric calculation: nickel(II) ions and dimethylglyoxime 201
3.5 Calculations associated with pharmaceuticals 202
3.6 Summary 202
3.7 End of topic test 203

Learning objective

By the end of this topic, you should be able to:

- carry out calculations from balanced equations (including multi-step reactions, reactant excess, and empirical formulae from given data);

- calculate the theoretical and actual yield from experimental data and provide explanations comparing the two.

3.1 Stoichiometry

Stoichiometry is the study of quantitative relationships involved in chemical reactions. The ability to balance and interpret equations is required to enable calculations to be carried out using any of the techniques covered in the practical skills and techniques topic.

3.2 Stoichiometric calculation: preparation of benzoic acid by hydrolysis of ethylbenzoate

This practical preparation of benzoic acid can be used to demonstrate the types of stoichiometric calculations that are required at Advanced Higher Chemistry. From the measurements obtained during the practical and from the balanced equation the percentage yield of benzoic acid can be calculated and compared with the theoretical yield which is calculated from the balanced equation.

Ethyl benzoate + NaOH → **Sodium benzoate** + $HOCH_2CH_3$

The first step involves the alkaline hydrolysis of the ester ethyl benzoate to form sodium benzoate.

Sodium benzoate (aq) + HCl(aq) → **Benzoic acid** (s) + Na^+Cl^-(aq)

The second step involves adding hydrochloric acid (strong acid) to precipitate out benzoic acid (weak acid). The crude benzoic acid is separated by filtration and recrystallised from water. The pure sample of benzoic acid is weighed and the percentage yield calculated. The melting point of the pure benzoic acid is determined.

The percentage yield of a product can be lowered due to many reasons including mass transfer of reactants/products, mechanical losses, purification of the product, side reactions, position of the equilibrium and the purity of the reactants.

TOPIC 3. STOICHIOMETRIC CALCULATIONS

Examples of results obtained from an experiment:

Mass	Weight
Mass of round bottomed flask	40.25 g
Mass of round bottomed flask + ethyl benzoate	45.61 g
Mass of ethyl benzoate	5.36 g
Mass of clock glass	9.62 g
Mass of clock glass + benzoic acid	12.86 g
Mass of benzoic acid	3.24 g

From the two-step balanced equations for this reaction it can be noted that 1 mole of ethyl benzoate is required to produce 1 mole of benzoic acid. From this we can work out the theoretical yield of benzoic acid.

$$1 \text{ mole of ethyl benzoate} \rightarrow 1 \text{ mole of benzoic acid}$$
$$150.0 \text{ g} \rightarrow 122.0 \text{ g}$$
$$5.36 \text{ g (quantity used in experiment)} \rightarrow (5.36 \times 122.0) / 150.0 = \mathbf{4.36 \text{ g}}$$

4.36 g is the theoretical yield of benzoic acid that will be produced from 5.36 g of ethyl benzoate according to the molar ratios of the balanced equations.

As reactions are rarely 100% effective due to the reasons given previously, we can calculate the actual percentage yield of benzoic acid by comparing the theoretical yield to the mass actually produced in the experiment.

Theoretical yield = 4.36 g
Actual yield = 3.24 g

Percentage (%) yield
= Actual yield/Theoretical yield × 100
= 3.24/4.36 × 100
= **74.3%**

Other experiments including the preparation of aspirin and potassium trioxalatoferrate (III) can be used in the same way to calculate the percentage yield of product by comparing the theoretical yield calculated from the balanced equations to the actual yield which is weighed at the end of the experiment.

3.3 Reactant in excess

In the reaction between calcium carbonate and hydrochloric acid they react according to the following equation:

$$CaCO_3 + 2HCl \rightarrow CaCl_2 + CO_2 + H_2O$$

The calcium carbonate and hydrochloric acid react in a molar ratio of 1:2 where 1 mole of calcium carbonate ($CaCO_3$) reacts with 2 moles of hydrochloric acid (HCl). In this case both reactants would be used up completely in the reaction. If there was a shortage of calcium carbonate for example then the reaction would stop when it runs out and there would be hydrochloric acid left over. In other words the hydrochloric acid would be in excess.

Excess calculation example

Calculate which reactant is in excess, when 10 g of calcium carbonate reacts with 50 cm^3 of 2 mol l^{-1} hydrochloric acid. This calculation involves the reactants only.

Firstly work out the number of moles of each reactant involved.

Calcium carbonate - number of moles
= mass/gfm
= 10/100
= **0.1 moles**

Hydrochloric acid - number of moles
= concentration × volume (litres)
= 2 × (50/1000)
= 2 × 0.05
= **0.1 moles**

From the balanced equation 1 mole of calcium carbonate reacts with 2 moles of hydrochloric acid and therefore 0.1 moles of calcium carbonate would react with 0.2 moles of hydrochloric acid.

From the previous calculation of the number of moles of hydrochloric acid you can see that there is only 0.1 moles available which is insufficient and therefore the calcium carbonate is in excess and some will be left over at the end of the reaction which will terminate once the hydrochloric acid has been used up.

Further examples of excess calculations are as follows.

Excess calculations   Go online

Q1: Calculate which reactant is in excess when 0.654 g of zinc reacts with 20 cm^3 of hydrochloric acid, concentration 0.5 mol l^{-1}, using the following equation:

$$Zn + 2HCl \rightarrow ZnCl_2 + H_2$$

a) Zinc
b) Hydrochloric acid

TOPIC 3. STOICHIOMETRIC CALCULATIONS

Q2: Calculate which reactant is in excess when 25 cm^3 of 1 mol l^{-1} sulfuric acid is mixed with 25 cm^3 of 1 mol l^{-1} sodium hydroxide, using the following equation:

$$H_2SO_4 + 2NaOH \rightarrow Na_2SO_4 + 2H_2O$$

a) Sulfuric acid
b) Sodium hydroxide

3.4 Stoichiometric calculation: nickel(II) ions and dimethylglyoxime

Nickel(II) ions and dimethylglyoxime   Go online

Nickel(II) ions react quantitatively with dimethylglyoxime ($C_4H_8O_2N_2$) forming a complex which precipitates out as a red solid. The gram formula mass of the complex is 288.7. The equation for the reaction and the structure of the complex are shown.

$$Ni^{2+} + 2C_4H_8O_2N_2 \rightarrow Ni(C_4H_7O_2N_2)_2 + 2H^+$$

Equation for nickel(II) ions and dimethylglyoxime reaction

Structure of nickel(II) dimethylglyoxime complex

Q3: What is the coordination number of nickel in the complex?

3.5 Calculations associated with pharmaceuticals

Percentage solution by mass is the mass of solute made up to 100 cm^3 of solution.

% mass = (mass of solute ÷ mass of solution) × 100

> **Example** 100 g of salt solution has 30 g of salt in it.
>
> % mass = (30 ÷ 100) × 100 = 30%

Percentage solution by volume is the number of cm^3 of solute made up to 100 cm^3 of solution.

% volume = (volume of solute ÷ volume of solution) × 100

> **Examples**
>
> 1. Wine with 12 ml alcohol per 100 ml of solution.
>
> % volume = (12 ÷ 100) × 100 = 12%
>
> ....
>
> 2. Making 1000 ml of 10% ethylene glycol solution.
>
> 10% = 0.1 × 1000 = 100 ml of ethylene glycol.
>
> The other 900 ml would be made up using water (although this would need to be topped up due to the two liquids not mixing to form exactly 1000 ml).

In other calculations, you may come across the unit *parts per million (ppm)* which refers to 1 mg per kg.

3.6 Summary

> **Summary**
>
> You should now be able to:
>
> - carry out calculations from balanced equations, including multi-step reactions, reactant excess, and empirical formulae from given data;
> - calculate the theoretical and actual yield from experimental data and provide explanations comparing the two.

3.7 End of topic test

End of Topic 3 test   Go online

Aspirin tablet back titration

The acetylsalicylic acid ($C_9H_8O_4$) content of an aspirin tablet was determined using a back titration.

Five aspirin tablets were crushed and added to 25.0 cm^3 of 1.00 mol l^{-1} sodium hydroxide solution. The mixture was heated and allowed to simmer for 30 minutes.

The resulting mixture (see previous diagram) was allowed to cool before being transferred to a 250 cm^3 standard flask and made up to the mark with deionised water. 25.0 cm^3 samples of this solution were titrated with 0.050 mol l^{-1} sulfuric acid, using the following equation:

$$H_2SO_4 + 2NaOH \rightarrow Na_2SO_4 + 2H_2O$$

Results of the titrations are included in the following table.

	Rough titration	First titration	Second titration
Initial burette reading/cm^3	0.0	9.0	17.7
Final burette reading/cm^3	9.0	17.7	26.3
Volume used/cm^3	9.0	8.7	8.6

Q4: Which of the following indicators would be the best one to use in the back titration? *(1 mark)*

a) screened methyl orange
b) bromocresol green
c) thymolphthalein
d) phenolphthalein

....

Q5: Calculate the number of moles of sulfuric acid in the average titre. *(2 marks)*

....

Q6: Calculate the number of moles of sodium hydroxide in the standard flask. *(2 marks)*

....

Q7: Calculate the number of moles of sodium hydroxide which reacted with the acetylsalicylic acid. *(2 marks)*

....

Q8: The mass of one mole of acetylsalicylic acid is 180 g. Use this along with the answer to the previous question to calculate the mass of acetylsalicylic acid in one aspirin tablet. *(3 marks)*

Q9: A compound contains only carbon, hydrogen and sulfur. Complete combustion of the compound gives 3.52 g of carbon dioxide, 2.16 g of water and 2.56 g of sulfur dioxide.

Show by calculation that the empirical formula is C_2H_6S. *(4 marks)*

Benzoic acid is prepared from ethyl benzoate by refluxing with sodium hydroxide solution (the following diagram shows a shortened version of the equation provided to carry out the calculation).

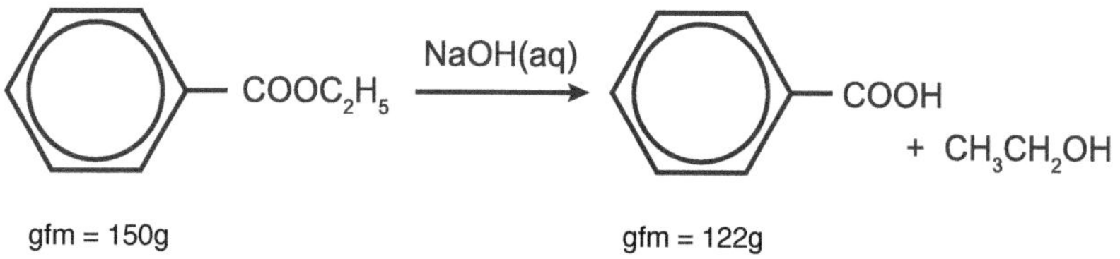

$$C_6H_5\text{-}COOC_2H_5 \xrightarrow{NaOH(aq)} C_6H_5\text{-}COOH + CH_3CH_2OH$$

gfm = 150g   gfm = 122g

A yield of 73.2% of benzoic acid was obtained from 5.64 g of ethyl benzoate.

Q10: Calculate the mass of benzoic acid produced. *(2 marks)*

Unit 3 Topic 4

Gravimetric analysis

Contents

4.1 Gravimetric analysis 206
4.2 Gravimetric analysis experiments 206
4.3 Summary 209

Prerequisites

Before you begin this topic, you should be able to work out the empirical formula from data given. (Advanced Higher Chemistry)

Learning objective

By the end of this topic, you should be able to:

- use an accurate electronic balance;
- explain what is meant by 'weighing by difference';
- explain what is meant by 'heating to constant mass'.

4.1 Gravimetric analysis

Measurements for gravimetric analysis are masses in grams and are typically determined using a digital balance that is accurate to two, three or even four decimal places. Gravimetric analysis is a quantitative determination of an analyte based on the mass of a solid.

The mass of the analyte present in a substance is determined by changing that chemical substance into solid by precipitation (with an appropriate reagent) of known chemical composition and formula. The precipitate needs to be readily isolated, purified and weighed. Often it is easier to remove the analyte by evaporation.

The final product has to be dried completely which is done by 'heating to constant mass'. This involves heating the substance, allowing it to cool in a desiccator (dry atmosphere) and then reweighing it all in a crucible. During heating the crucible lid should be left partially off to allow the water to escape. A blue flame should be used for heating to avoid a build up of soot on the outside of the crucible which could affect the mass. Heating should be started off gently and then more strongly.

The mass of the crucible is measured before adding the substance and the final mass of the substance is determined by subtracting the mass of the crucible from the mass of the crucible and dried substance (weighing by difference). This process of heating, drying and weighing is repeated until a constant mass is obtained. This shows that all the water has been driven off.

4.2 Gravimetric analysis experiments

Experiment 1: Gravimetric analysis of water in hydrated barium chloride

This experiment determines the value of n in the formula $BaCl_2.nH_2O$ using gravimetric analysis through accurate weighing. The hydrated barium chloride is heated until all the water has been removed as in the equation:

$$BaCl_2.nH_2O \rightarrow BaCl_2 + nH_2O$$

From the masses measured in the experiment it is possible to calculate the relative number of moles of barium chloride and water and hence calculate the value of n (must be a whole number).

The experiment is carried out in a crucible which is heated first to remove any residual water, cooled in a desiccator and then weighed accurately. Approximately 2.5 g of hydrated barium chloride is then added to the crucible and again weighed accurately (crucible plus contents). The hydrated barium chloride is heated in the crucible using a blue Bunsen flame after which the crucible and contents are placed in a desiccator to cool (desiccator contains a drying agent which reduces the chances of any moisture from the air being absorbed by the sample). After cooling the crucible and contents are reweighed accurately.

The heating, cooling and reweighing is repeated until a constant mass is achieved and the assumption that all the water has been removed.

Example of results obtained

Mass of empty crucible = 32.67 g
Mass of crucible + hydrated barium chloride = 35.03 g
Mass of hydrated barium chloride = 35.03 - 32.67 = 2.36 g

Mass of crucible and anhydrous barium chloride = 34.69 g (heated to constant mass)
Mass of anhydrous barium chloride = 34.69 - 32.67 = 2.02 g

Mass of water removed = 2.36 - 2.02 = 0.34 g
Moles of water removed = 0.34 / 18 = 0.0189

Number of moles of barium chloride $BaCl_2$ = 2.02 / 208.3 = 0.00970

Ratio of moles = 0.00970:0.0189 = $BaCl_2$:H_2O = 1:2

Formula of $BaCl_2.nH_2O$ is therefore $BaCl_2.2H_2O$ where n=2.

Experiment 2: Gravimetric determination of water in hydrated magnesium sulfate

As for the gravimetric determination of water in barium chloride the water in magnesium sulfate ($MgSO_4.nH_2O$) can also be determined using this technique. The same experimental procedure is used substituting hydrated barium chloride for hydrated magnesium sulfate.

Experiment 3: Determination of nickel using butanedioxime (dimethylglyoxime)

Dimethylgloxime is used as a chelating agent in the gravimetric analysis of nickel. It has the formula $CH_3C(NOH)C(NOH)CH_3$.

Dimethylglyoxime

Dimethylglyoxime   *Insoluble complex*

The nickel is precipitated as red nickel dimethylglyoxime by adding an alcoholic solution of dimethylglyoxime and then adding a slight excess of aqueous ammonia solution. When the pH is buffered in the range of 5 to 9, the formation of the red chelate occurs quantitatively in a solution. The chelation reaction occurs due to donation of the electron pairs on the four nitrogen atoms, not by electrons on the oxygen atoms. The reaction is performed in a solution buffered by either an ammonia or citrate buffer to prevent the pH of the solution from falling below 5.

The mass of the nickel is determined from the mass of the precipitate which is filtered, washed and dried to constant mass as in the other examples.

Mass of crucible + lid = w1 (w=weight)
Mass of crucible + lid + dried precipitate = w2
Mass of precipitate = w2 - w1 = w3

GFM of precipitate $Ni(C_4H_7O_2N_2)_2$ = 288.7 g
GFM of nickel (Ni) = 58.7 g

Mass of nickel = (w3 × 58.7) / 288.7 g

Gravimetric analysis Go online

Q1: Describe the process of heating to constant mass.

....

Q2: During the process of heating to constant mass a desiccator is used. Explain why a desiccator is used.

....

Q3: Explain why gravimetric analysis is a suitable technique for determining nickel with dimethylglyoxime.

....

Q4: When 0·968 g of an impure sample of nickel(II) sulfate, $NiSO_4.7H_2O$ was dissolved in water and reacted with dimethylglyoxime, 0·942 g of the red precipitate was formed.

Calculate the percentage of nickel in the impure sample.

TOPIC 4. GRAVIMETRIC ANALYSIS

4.3 Summary

Summary

You should now be able to:

- use an accurate electronic balance;
- explain what is meant by 'weighing by difference';
- explain what is meant by 'heating to constant mass'.

Unit 3 Topic 5

Volumetric analysis

Contents

- 5.1 Volumetric analysis 213
- 5.2 Preparation of a standard solution 214
- 5.3 Use of controls in chemical reactions 216
- 5.4 Complexometric titrations 216
- 5.5 Back titrations and associated calculations 218
- 5.6 Summary 220
- 5.7 Resources 220
- 5.8 End of topic test 220

Prerequisites

Before you begin this topic, you should know:

- about the key area of chemical analysis; (National 5 Chemistry)
- how and when to use pipettes, burettes and standard/volumetric flasks (National 5 and Higher Chemistry);
- about ligands/complexes and indicator/choice of indicator (Advanced Higher Chemistry).

Learning objective

By the end of this topic you should:

- be able to prepare a standard solution;
- understand that standard solutions can be prepared from primary standards;
- be able to state that a primary standard must have the following characteristics:
    - a high state of purity;
    - stability in air and water;
    - solubility;
    - reasonably high formula mass;
- understand the role of a control in experiments to validate a technique;
- understand the use of complexometric titrations in quantitative analysis of solutions containing a metal ion;
- understand the use of back titrations;
- understand how to use calculations associated with back titrations.

TOPIC 5. VOLUMETRIC ANALYSIS

5.1 Volumetric analysis

Volumetric analysis involves using a solution of known concentration (standard solution) in a quantitative reaction to determine the concentration of the other reactant. The procedure used to carry out volumetric analysis is titration whether in the form of standard, complexometric or back titrations.

Titrations involve measuring one solution quantitatively into a conical flask using a pipette. The other solution is added from a burette until a permanent colour change of an indicator is seen in the conical flask.

A 'rough' titration is carried out first followed by more accurate titrations until concordant titre values are achieved (titre volumes added from burette should be ± 0.1 cm^3 of each other). The mean or average value of the concordant titres is used in calculations.

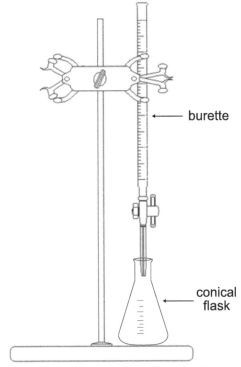

Titration using a burette and conical flask

5.2 Preparation of a standard solution

A standard solution is one of which the concentration is known accurately and can be prepared directly from a primary standard.

A primary standard must have, at least, the following characteristics:

- a high state of purity;
- stability in air and water;
- solubility;
- reasonably high formula mass;

The standard solution is prepared as follows:

- calculate the mass of the primary standard required to make the concentration of solution required in the appropriate volume of solution;
- weigh out the primary standard as accurately as possible;
- dissolve the primary standard in a small volume of deionised water in a beaker;
- transfer the solution and all the rinsings into a standard flask;
- make the solution up to the mark with more deionised water;
- invert the stoppered standard flask several times to ensure thorough mixing.

A standard (volumetric) flask

Substances that are used as primary standards include:

- Oxalic acid ($H_2C_2O_4.2H_2O$);
- Sodium carbonate (Na_2CO_3);
- Potassium hydrogen phthalate ($KH(C_8H_4O_4)$);
- Potassium iodate (KIO_3);
- Potassium dichromate ($K_2Cr_2O_7$).

It is not possible to use substances such as sodium hydroxide (NaOH) as primary standards as they readily absorb water and carbon dioxide from the atmosphere.

TOPIC 5. VOLUMETRIC ANALYSIS

Experiment: Preparation of 0.1 mol l⁻¹ oxalic acid

Oxalic acid $H_2C_2O_4.2H_2O$ formula mass = 126 g

In 1000 cm^3 (1 litre) you would need to weigh out **12.6 g** (0.1 × 126) which would make a solution with a concentration of 0.1 mol l^{-1}.

In 250 cm^3 you would need to weigh out $n = c \times v$ **moles** of oxalic acid to make a 0.1 mol l^{-1} solution.

n = 0.1 × 0.25 moles = 0.025 moles

0.025 moles of oxalic acid = ***0.025 × 126 = 3.15 g***

Other possible experiments include:

a) Standardisation of approximately 0.1 mol l^{-1} NaOH
b) Determination of ethanoic acid content of vinegar
c) Preparation of a standard solution of 0.1 mol l^{-1} Na$_2$CO$_3$
d) Standardisation of approximately 0.1 mol l^{-1} HCl
e) Determination of purity of marble by back titration (see section 2.5)

Standardisation of approximately 0.1 mol l⁻¹ NaOH/0.1 mol l⁻¹ HCl

Standardisation is the process of determining the exact concentration of a solution. Titration is one type of analytical procedure often used in standardisation.

The point at which the reaction is complete in a titration is referred to as the endpoint. A chemical substance known as an indicator is used to indicate the endpoint. The indicator used in this experiment is phenolphthalein. Phenolphthalein, an organic compound, is colourless in acidic solution and pink in basic solution.

This experiment involves two separate acid-base standardisation procedures. In the first standardisation the concentration of a sodium hydroxide solution (NaOH) will be determined by titrating a sample of potassium acid phthalate (KHP; HKC$_8$H$_4$O$_4$) with the NaOH. This is the primary standard.

Experiment: first procedure

A 0.128 g sample of potassium acid phthalate (KHP, HKC$_8$H$_4$O$_4$) required 28.5 cm^3 of NaOH solution to reach a phenolphthalein endpoint. Calculate the concentration of the NaOH.

$$HKC_8H_4O_4 \rightarrow NaKC_8H_4O_4 + H_2O$$

Moles of KHP (mass / gfm) = 0.128 / 204.1 = 6.271 × 10^{-4} mol

Moles of NaOH (KHP:NaOH = 1:1) = 6.271 × 10^{-4} mol

Concentration of NaOH (moles / volume in litres)
= 6.271 × 10^{-4} / 0.0285 = ***0.0220 mol l⁻¹ NaOH***

© HERIOT-WATT UNIVERSITY

In the second procedure the standardised NaOH will be used to determine the concentration of a hydrochloric solution (HCl).

> **Experiment: second procedure**
>
> A 20.00 cm^3 sample of HCl was titrated with the NaOH solution from experiment 1. To reach the endpoint required 23.7 cm^3 of the NaOH. Calculate the concentration of the HCl.
>
> $$HCl + NaOH \rightarrow NaCl + H_2O$$
>
> Moles of NaOH (concentration × volume) = 0.0237 × 0.0220 = 5.214 × 10^{-4} mol
>
> Moles of HCl (HCl:NaOH = 1:1) = 5.214 × 10^{-4} mol
>
> Concentration of HCl (moles / volumes in litres)
> = 5.214 × 10^{-4} / 0.02000 = ***0.0261 mol l^{-1}***

5.3 Use of controls in chemical reactions

The use of a control in chemical reactions validates a technique and may consist of carrying out a determination on a solution of known concentration.

In the determination of the percentage of acetyl salicylic acid in commercial aspirin tablets a sample of pure aspirin (100% aspirin) would also be analysed to validate the techniques being used. In the determination of the vitamin C content in fruit juice a sample of ascorbic acid (pure vitamin C) would also be analysed again to validate the techniques being used in the determination and to give a referencing point on which to base all other results from impure samples.

5.4 Complexometric titrations

Complexometric titration is a form of volumetric analysis in which the formation of a coloured complex is used to indicate the end point of a titration. Complexometric titrations are particularly useful for the determination of a mixture of different metal ions in solution.

EDTA (ethylenediaminetetraacetic acid) is a hexadentate ligand and an important complexometric reagent used to determine the concentration of metal ions in solution forming an octahedral complex with the metal 2+ ion in a 1:1 ratio. In particular it can be used to determine the concentration of nickel ions in a nickel salt. To carry out metal ion titrations using EDTA, it is almost always necessary to use a complexometric indicator to determine when the end point has been reached.

Murexide is used and, compared to its colour when it is attached to the Ni^{2+} ions, is a different colour when free. Murexide is a suitable indicator as it binds less strongly to the Ni^{2+} ions than the EDTA does and so is no longer attached to the Ni^{2+} ions at the end point of the titration where the colour is changed from yellow to blue-purple. Titrations would be carried out until concordant results were obtained.

TOPIC 5. VOLUMETRIC ANALYSIS

In this experiment ammonium chloride and ammonia solutions are used as a buffer to keep the pH constant as murexide is a pH dependent indicator.

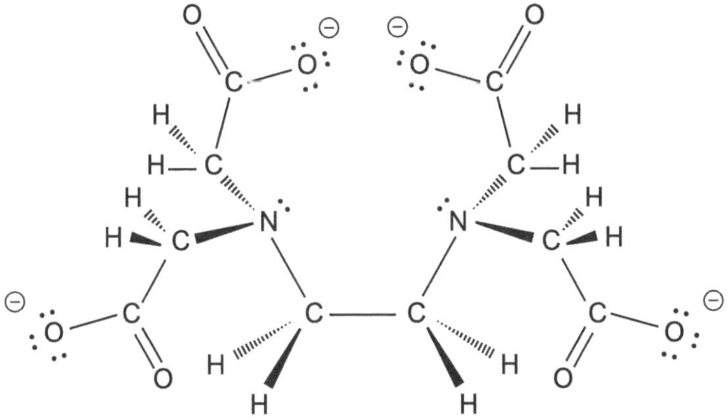

EDTA

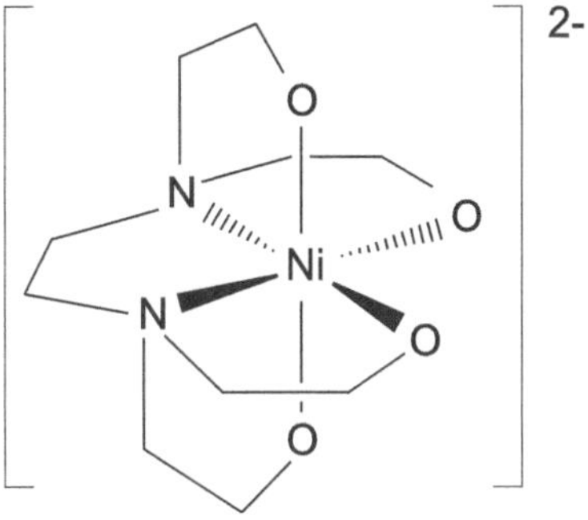

EDTA complexed with nickel

Worked example for Ni-EDTA titration

Approximately 2.6 g of hydrated nickel sulfate ($NiSO_4.6H_2O$) was weighed accurately and dissolved in a small volume of deionised water, transferred to a 100 cm^3 standard flask and made up to the mark with deionised water. 20cm^3 samples of this solution were titrated with 0.112 mol l^{-1} EDTA solution using murexide as an indicator. The following results were obtained.

Mass of weighing boat + $NiSO_4.6H_2O$ = 4.076 g

Mass of weighing boat after transferring $NiSO_4.6H_2O$ = 1.472 g

Titre	Initial burette reading (cm³)	Final burette reading (cm³)	Volume of EDTA added(cm³)
1	1.4	20.5	19.1
2	20.5	38.6	18.1
3	15.3	33.5	18.2

Calculate the % of nickel in the hydrated nickel sulfate.

Average titre volume = 18.15 cm³

Number of moles of EDTA used = C × V = 0.112 × 0.01815 = 0.00203 mol

Since EDTA complexes with nickel ions in a 1:1 ratio, the number of moles of Ni^{2+} in 20 cm³ = 0.00203 mol

Therefore, in 100 cm³, there are 0.00203 × 5 = 0.0102 mol

Mass of nickel present = n × GFM = 0.0102 × 58.7 = 0.599 g

Mass of $NiSO_4.6H_2O$ = 4.076 - 1.472 = 2.604 g

% nickel = (0.599/2.604) × 100 = 23.0 %

5.5 Back titrations and associated calculations

A back titration is used to find the number of moles of a substance by reacting it with an excess volume of reactant of known concentration. The resulting mixture is then titrated to work out the number of moles of the reactant in excess. A back titration is useful when trying to work out the quantity of substance in an insoluble solid.

From the initial number of moles of that reactant the number of moles used in the reaction can be determined, making it possible to work back to calculate the initial number of moles of the substance under test.

An experiment which uses a back titration is in the determination of aspirin due to it being insoluble in water. A sample of aspirin of accurately known mass is treated with an excess of sodium hydroxide (the actual volume is known). The sodium hydroxide catalyses the hydrolysis of aspirin to ethanoic acid and salicylic acid and then neutralises these two acids.

Determination of aspirin

TOPIC 5. VOLUMETRIC ANALYSIS

As an excess of sodium hydroxide is used the volume remaining is determined by titrating it against a standard solution of sulfuric acid. The difference between the initial and excess volumes of sodium hydroxide allows the mass of aspirin in the tablet to be determined.

The reaction taking place is:

$$2NaOH + H_2SO_4 \rightarrow Na_2SO_4 + 2H_2O$$

Determination of aspirin

Three aspirin tablets of approximately 1.5 g were added to a conical flask. Sodium hydroxide (25.0 cm^3 1.00 mol l^{-1}) was pipetted into the flask along with 25.0 cm^3 water. The resulting mixture was simmered gently on a hot plate for approximately 30 minutes and after cooling transferred along with rinsings to a 250 cm^3 standard flask. The solution was made up to the graduation mark with water, stoppered and inverted several times to ensure through mixing of the contents. 25.0 cm^3 was pipetted into a conical flask; phenolphthalein indicator added and titrated against sulfuric acid (0.05 mol l^{-1}). The end point of the titration was indicated by the colour change pink to colourless and they were repeated until concordant results were obtained ($\pm$0.1 cm^3).

Titre	Starting volume cm^3	End volume cm^3	Titre volume cm^3
Rough	0.0	15.6	15.6
1	15.6	30.8	15.2
2	0.0	15.1	15.1

Titration results

Average titre volume = (15.1 + 15.2) / 2 = 15.15 cm^3

Number of moles of sulfuric acid used = 0.01515 $\times$ 0.05 = 7.575 x 10^{-4} mol

Number of moles of NaOH left in the 25.0 cm^3 hydrolysed solution = 2 $\times$ 7.575 $\times$ 10^{-4} = 1.515 $\times$ 10^{-3} mol

Number of moles of NaOH left in 250.0 cm^3 of the hydrolysed solution = 10 $\times$ 1.515 $\times$ 10^{-3} = 1.515 $\times$ 10^{-2} mol

Number of moles of NaOH added to aspirin initially = 0.025 $\times$ 1 = 2.5 x10^{-2} mol

Number of moles NaOH reacted with aspirin = 2.5 $\times$ 10^{-2} - 1.515 $\times$ 10^{-2} = 9.85 $\times$ 10^{-3} mol

2 moles of NaOH reacts with 1 mole aspirin therefore number of moles of aspirin in 3 tablets = 9.85 $\times$ 10^{-3} / 2 = 4.925 $\times$ 10^{-3} mol

Number of moles of aspirin in 1 tablet = 4.925 $\times$ 10^{-3} / 3 = 1.642 $\times$ 10^{-3} mol

Mass of aspirin in each tablet = n $\times$ gfm = 1.642 $\times$ 10^{-3} x 180 = 0.296 g = 296 mg

5.6 Summary

Summary

You should now:

- be able to prepare a standard solution;
- understand that standard solutions can be prepared from primary standards;
- be able to state that a primary standard must have the following characteristics:
    - a high state of purity;
    - stability in air and water;
    - solubility;
    - reasonably high formula mass;
- understand the role of a control in experiments to validate a technique;
- understand the use of complexometric titrations in quantitative analysis of solutions containing a metal ion;
- understand the use of back titrations;
- understand how to use calculations associated with back titrations.

5.7 Resources

- Determination of the molarity of an acid or base solution (http://bit.ly/2sK5SV7)
- Titration experiment (http://bit.ly/1ucRlHo)

5.8 End of topic test

End of Topic 5 test   Go online

Q1: The equation for the reaction between benzoic acid solution and sodium hydroxide solution is:

$$C_6H_5COOH(aq) + NaOH(aq) \rightarrow C_6H_5COONa(aq) + H_2O(l)$$

A student used a standard solution of 0.0563 mol l^{-1} benzoic acid to standardise 20.0 cm^3 of approximately 0.05 mol l^{-1} sodium hydroxide solution.

The results for the titration are given in the following table.

TOPIC 5. VOLUMETRIC ANALYSIS

	1st attempt	2nd attempt	3rd attempt
Final burette reading/cm^3	17.2	33.8	16.6
Initial burette reading/cm^3	0.0	17.2	0.1
Titre/cm^3	17.2	16.6	16.5

Calculate the accurate concentration of the sodium hydroxide solution (to four decimal places). *(3 marks)*

....

Q2: Four aspirin tablets of approximately 2.0 g were added to a conical flask. Sodium hydroxide (25.0 cm^3 1.00 mol l^{-1}) was pipetted into the flask along with 25.0 cm^3 water. The resulting mixture was simmered gently on a hot plate for approximately 30 minutes and after cooling transferred along with rinsings to a 250 cm^3 standard flask. The solution was made up to the graduation mark with water, stoppered and inverted several times to ensure through mixing of the contents. 25.0 cm^3 was pipetted into a conical flask; phenolphthalein indicator added and titrated against sulfuric acid (0.02 mol l^{-1}). The end point of the titration was indicated by the colour change pink to colourless and they were repeated until concordant results were obtained (±0.1 cm^3).

+ 2NaOH → + H$_2$O

$$2NaOH + H_2SO_4 \rightarrow Na_2SO_4 + 2H_2O$$

Titre	Starting volume cm^3	End volume cm^3	Titre volume cm^3
Rough	0	14.9	14.9
1	14.9	30.8	15.9
2	0	15.8	15.8

Titration results

Calculate the mass of aspirin in each tablet in grams. *(3 marks)*

Unit 3 Topic 6

Practical skills and techniques

Contents

6.1	Colorimetry and accurate dilution	224
6.2	Distillation	225
6.3	Refluxing	226
6.4	Vacuum filtration	227
6.5	Recrystallisation	229
6.6	Use of a separating funnel	229
6.7	Thin-layer chromatography (TLC)	230
6.8	Melting point and mixed melting point	233
6.9	Summary	234
6.10	Resources	235
6.11	End of topic test	235

Learning objective

By the end of this topic, you should be able to:

- prepare standard solutions using accurate dilution technique;
- form and use calibration curves to determine an unknown concentration using solutions of appropriate concentration;
- gain knowledge of the appropriate use of distillation, reflux, vacuum filtration, recrystallisation and use of a separating funnel in the preparation and purification of an experimental product;
- gain knowledge of the appropriate uses of thin-layer chromatography, melting point and mixed melting point determination in evaluating the purity of experimental products;
- calculate Rf values from relevant data and their use in following the course of a reaction.

6.1 Colorimetry and accurate dilution

Colorimetry uses the relationship between colour intensity of a solution and the concentration of the coloured species present to determine the concentration. A colorimeter consists essentially of a light source, a coloured filter, a light detector and a recorder. The filter colour is chosen as the complementary colour to that of the solution resulting in maximum absorbance. The light passes through the filter and then through the coloured solution and the difference in absorbance between the coloured solution and water is detected and noted as an absorbance value.

Colorimetry is a useful method of analysis for coloured compounds since at low concentrations, coloured solutions obey the Beer-Lambert law which is usually quoted as $A = \varepsilon c l$ where:

- ε = molar absorptivity (this is a constant dependent on the solution at a particular wavelength);
- c = the concentration of solution;
- l = path length (the distance the light travels through the sample solution).

The absorbance of the solution is calculated using $A = \log(I_0/I)$ where I_0 is the initial intensity of light before passing through the solution and I is the intensity of light after passing through the solution.

A calibration curve must be prepared using solutions of known concentrations (standard solutions). The unknown concentration of the solution is determined from its absorbance and by referring to the calibration curve. The straight line section of the calibration graph should cover the dilution range likely to be used in the determination.

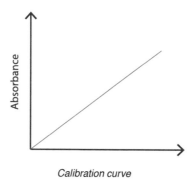

Calibration curve

Making of standard solutions:

a) Dissolve the accurately weighed substance in a small volume of water in a beaker.

b) Pour the solution into a standard flask.

c) Rinse the beaker with distilled water and add the rinsings to the standard flask.

d) Add distilled water to the standard flask making the volume up to the mark.

e) Stopper the flask and invert several times to ensure thorough mixing.

Experiments including the determination of manganese in steel (see Advanced Higher Chemistry. A Practical Guide Support Materials (https://bit.ly/2n4gpW5)) and determination of nickel could both be used to gain skills in this practical technique.

Document

- Determination of manganese in steel by permanganate colorimetry (https://bit.ly/2MbMTsy)

6.2 Distillation

In the modern organic chemistry laboratory, distillation is a powerful tool, both for the identification and the purification of organic compounds. The boiling point of a compound which can be determined by distillation is well-defined and thus is one of the physical properties of a compound by which it is identified.

Distillation is used to purify a compound by separating it from a non-volatile or less-volatile material.

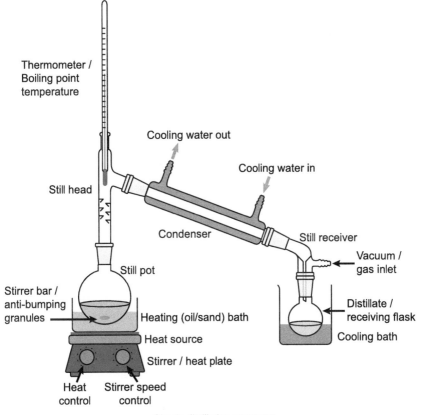

Simple distillation apparatus

Steam distillation is a special type of distillation for temperature sensitive materials like natural aromatic compounds.

Fractional distillation is the separation of a mixture into its component parts, or fractions, such as in separating chemical compounds by their boiling point by heating them to a temperature at which one or more fractions of the compound will vaporise. Used in the oil industry in the separation of crude oil into fractions that are used mostly as fuels.

Possible experiments for distillation include:

- Preparation of benzoic acid by hydrolysis of ethyl benzoate;
- Preparation of ethyl ethanoate;
- Preparation of cyclohexene from cyclohexanol.

All of these experiments are shown in the practical guide for revised Advanced Higher Chemistry (see Advanced Higher Chemistry. A Practical Guide Support Materials (https://bit.ly/2n4gpW5) .

Videos

- Steam distillation (http://bit.ly/1neY5aK)
- Fractional distillation (http://rsc.li/2bVREXX)
- Preparation of ethyl ethanoate (http://bit.ly/1PbLXO5)

6.3 Refluxing

Refluxing is a technique used to apply heat energy to a chemical reaction mixture over an extended period of time. The liquid reaction mixture is placed in a round-bottomed flask along with anti-bumping granules with a condenser connected at the top. The flask is heated vigorously over the course of the chemical reaction; any vapours given off are immediately returned to the reaction vessel as liquids when they reach the condenser.

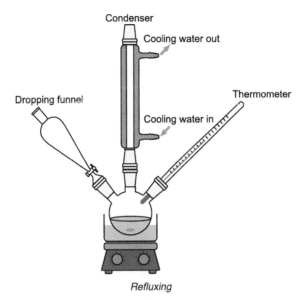

Refluxing

Possible experiments to use this technique include:

- Preparation of benzoic acid by hydrolysis of ethyl benzoate;
- Preparation of ethyl ethanoate.

Both of these experiments are shown in the practical guide for revised Advanced Higher Chemistry (see Advanced Higher Chemistry. A Practical Guide Support Materials (https://bit.ly/2n4gpW5) .

6.4 Vacuum filtration

Vacuum filtration can be carried out using a Buchner, Hirsch or sintered glass funnel. These methods are carried out under reduced pressure and provide a faster means of separating the precipitate from the filtrate. The choice of filtering medium depends on the quantity and nature of the precipitate.

Possible experiments include:

- Preparation of potassium trioxolatoferrate(III);
- Preparation of aspirin;
- Preparation of benzoic acid by hydrolysis of ethyl benzoate;
- Identification by derivative formation.

Hirsch funnel

Buchner flask

Sinter funnel

Buchner funnel

Hirsch funnels are generally used for smaller quantities of material. On top of the funnel part of both the Hirsch and Buchner funnels there is a cylinder with a fritted glass disc/perforated plate separating it from the funnel. A funnel with a fritted glass disc can be used immediately. For a funnel with a perforated plate, filtration material in the form of filter paper is placed on the plate, and the filter paper is moistened with a solvent to prevent initial leakage. The liquid to be filtered is poured into the cylinder and drawn through the perforated plate/fritted glass disc by vacuum suction.

Hot filtration method is mainly used to separate solids from a hot solution. This is done in order to prevent crystal formation in the filter funnel and other apparatuses that comes in contact with the solution. As a result, the apparatus and the solution used are heated in order to prevent the rapid decrease in temperature which in turn, would lead to the crystallization of the solids in the funnel and hinder the filtration process.

Videos

- How to use a Buchner funnel - http://bit.ly/1TV62wb
- Vacuum filtration - http://bit.ly/22ZuoLm

6.5 Recrystallisation

Recrystallisation is a laboratory technique used to purify solids, based upon solubility. The solvent for recrystallisation must be carefully selected such that the impure compound is insoluble at lower temperatures, yet completely soluble at higher temperatures. The impure compound is dissolved gently in the minimum volume of hot solvent then filtered to remove insoluble impurities. The filtrate is allowed to cool slowly to force crystallisation. The more soluble impurities are left behind in the solvent.

Recrystallisation can also be achieved where the pure compound is soluble in the hot solvent but not the cold solvent and the impurities are soluble in the hot and cold solvent. The impure compound is dissolved in the hot solvent. As it cools down the impurities stay in the solvent and can be filtered off.

Selection of a suitable solvent is crucial to achieve a satisfactory recrystallisation. Neither the compound nor the impurities should react with the solvent. Other factors to consider will be the solubility of the compound and the impurities in the solvent, and the boiling point of the solvent (if the boiling point is too low, it will not be possible to heat the solvent to dissolve the impurities as it will evaporate off too easily).

Possible experiments include:

- Preparation of benzoic acid by hydrolysis of ethyl benzoate;
- Preparation of potassium trioxalatoferrate(III);
- Preparation of acetylsalicylic acid.

All of these experiments are shown in the practical guide for revised Advanced Higher Chemistry (see Advanced Higher Chemistry. A Practical Guide Support Materials (https://bit.ly/2n4gpW5) .

6.6 Use of a separating funnel

Solvent extraction can be an application of the partition of a solute between two liquids. It is based on the relative solubility of a compound in two different immiscible liquids, usually water and an organic solvent. The partition coefficient is expressed as the concentration of a solute in the organic layer over that in the aqueous layer. The two solvents form two separate layers in the separating funnel and the lower layer is run off into one container and the upper layer is poured out into another container. The quantity of solute extracted depends on the partition coefficient and on the number of times that the process is repeated.

Again, selection of a suitable solvent is crucial to achieve a high concentration of the solute after the extraction. The solute should be more soluble in the solvent than in the aqueous solution and it should not react with the solvent.

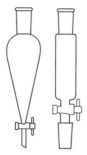

Separating funnels

Possible experiments include:

- Preparation of ethyl ethanoate;
- Extraction of caffeine from tea.

The largest risk when using a separating funnel is that of pressure build-up. Pressure accumulates during mixing if gas evolving reactions occur. This problem can be easily handled by simply opening the stopper at the top of the funnel routinely while mixing. This should be done with the top of the funnel pointed away from the body. When shaking, hold the stopper in place or it can become dislodged causing the liquids will spill. To account for this, simply hold the stopper in place with one hand.

Videos

- Liquid-liquid extraction using a separating funnel - http://bit.ly/1OPnUZo
- Industrial extraction of caffeine - http://bit.ly/22ZuvXg

6.7 Thin-layer chromatography (TLC)

Thin-layer chromatography can be used to assess product purity. Instead of chromatography paper, thin-layer chromatography (TLC) uses a fine film of silica or aluminium oxide spread over glass or plastic.

When setting up a TLC plate, a pencil line is drawn, usually 1 cm up from the bottom of the plate. The sample solution is then spotted on several times to get a concentrated spot on the plate. The plate is then placed in a suitable solvent (which acts as the mobile phase) making sure that the solvent is below the level of the spot. The solvent then travels up the plate and the components of the sample separate out according to their relative attractions to the stationary phase on the plate and the mobile phase (see the section on chromatography in 'Chemical equilibrium').

Retardation factor (R_f) values (distance travelled by compound/distance travelled by solvent) can be calculated and under similar conditions a compound will always have the same R_f value within experimental error. R_f values can be used to follow the course of a reaction by spotting a TLC plate with the authentic product, the authentic reactant and the reaction mixture at that point. Comparison of the R_f values of all three spots will allow the progress of the reaction to be determined.

Since a pure substance will show up as only one spot on the developed chromatogram, TLC can be used to assess the purity of a product prepared in the lab.

Possible experiments include:

- Preparation of aspirin;
- Hydrolysis of ethyl benzoate.

Chromatogram example

An organic chemist is trying to synthesis a fragrance compound by reacting compound X and compound Y together. After an hour of reaction, a sample is removed and compared with pure samples of X and Y using TLC. If the experiment produces pure product, then only one spot will be seen for the product sample. If impurities are present they will show up as additional spots on the chromatogram.

In the following chromatogram you can see that there are two spots produced from the product sample which have travelled the same distance as the spots for both compounds X and Y. This means the product sample only contains both compounds X and Y suggesting the chemical reaction has not taken place between X and Y.

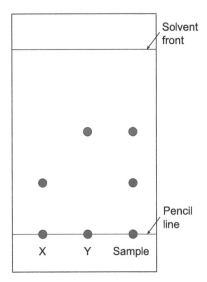

In the following chromatogram you can again see two spots in the product sample. This time, one of the spots has travelled further and is not in the same position as the spots for either compound X and compound Y. This suggests this spot is for the fragrance compound. The product sample still contains a spot at the same distance for compound X suggesting this is still present as an impurity.

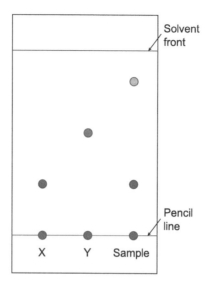

In the following chromatogram, only one spot is seen for the product sample which has travelled a different distance from the spots for both compound X and compound Y. This suggests the fragrance compound is pure.

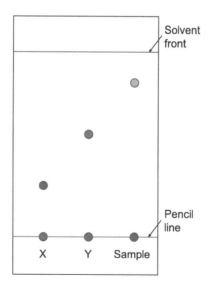

To check if a sample is pure, a co-spot can also be used. A 50:50 mixture of the sample and a known pure sample can be spotted on. If the sample is pure, then only one spot will be present on the chromatogram. If the sample is impure then more than one spot will be observed.

Measuring R_f values

R_f is measured by calculating the distance travelled by the substance divided by the distance travelled by the solvent front, as illustrated in the following diagram.

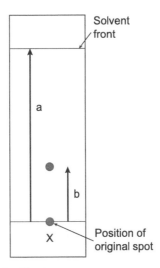

a = distance travelled by the solvent front

b = distance travelled by the substance

$R_f = b/a$

Videos

- Thin-layer chromatography (TLC) - http://bit.ly/1niBs5q
- Thin-layer chromatography (TLC) theory - http://bit.ly/1mW7x2n

6.8 Melting point and mixed melting point

The melting point of an organic compound is one of several physical properties by which it can be identified. A crystalline substance has a sharp melting point falling within a very small temperature range.

Determination of the melting point can also give an indication of the purity of an organic compound, as the presence of impurities lowers the melting point and extends its melting temperature range. Since impurities lower the melting point, the technique of mixed melting point determination can be used as a means of identifying the product of a reaction.

In this case the product can be mixed with a pure sample of the substance and a melting point taken of the mixture. If the melting point range is lowered and widened, it means that the two are different compounds. If the melting point stays the same it means that the two compounds are likely identical.

The melting point of an organic solid can be determined by introducing a tiny amount into a small capillary tube and placing inside the melting point apparatus. A window in the apparatus allows you to determine when the sample melts in the capillary tube and the temperature can be determined from the thermometer. Pure samples usually have sharp melting points, for example 149.5 - 150 °C or 189 - 190 °C; impure samples of the same compounds melt at lower temperatures and over a wider range, for example 145 - 148 °C or 186 - 189 °C.

Possible experiments include:

- Preparation of benzoic acid by hydrolysis of ethyl benzoate.
- Identification by derivative formation.
- Preparation of aspirin.

'Preparation of benzoic acid' and 'Preparation of aspirin' are shown in the following practical guide for Advanced Higher Chemistry (see Advanced Higher Chemistry. A Practical Guide Support Materials (https://bit.ly/2n4gpW5) .

6.9 Summary

Summary

You should now be able to:

- prepare standard solutions using accurate dilution technique;
- form and use calibration curves to determine an unknown concentration using solutions of appropriate concentration;
- gain knowledge of the appropriate use of distillation, reflux, vacuum filtration, recrystallisation and use of a separating funnel in the preparation and purification of an experimental product;
- gain knowledge of the appropriate uses of thin-layer chromatography, melting point and mixed melting point determination in evaluating the purity of experimental products;
- calculate Rf values from relevant data and their use in following the course of a reaction.

TOPIC 6. PRACTICAL SKILLS AND TECHNIQUES

6.10 Resources

- Determination of manganese in steel by permanganate colorimetry (https://bit.ly/2MbMTsy)
- Fractional distillation (http://rsc.li/2bVREXX)
- Steam distillation (http://bit.ly/1neY5aK)
- Advanced Higher Chemistry. A Practical Guide Support Materials (https://bit.ly/2n4gpW5)
- Preparation of ethyl ethanoate (http://bit.ly/1PbLXO5)
- LearnChemistry, Aspirin screen experiment (http://rsc.li/1EZAhew)
- How to use a Buchner funnel (http://bit.ly/1TV62wb)
- Vacuum filtration (http://bit.ly/22ZuoLm)
- Using a separating funnel (http://bit.ly/1OPnUZo)
- Thin-layer chromatography (TLC) (http://bit.ly/1niBs5q)
- Thin-layer chromatography (TLC) theory (http://bit.ly/1mW7x2n)

6.11 End of topic test

End of Topic 6 test   Go online

Q1: Small amounts of manganese are added to aluminium in the making of drink cans to prevent corrosion. Using colorimetry the concentration of manganese in the alloy can be determined through the conversion of manganese to permanganate ions.
Describe how this is done. *(3 marks)*

....

Q2: A student was measuring the percentage of calcium carbonate in different types of egg shell. The egg shells were ground and approximately 0.5 g were weighed accurately of each.
What is meant by weighing accurately approximately 0.5 g? *(1 mark)*

....

Q3: The egg shells were placed in a beaker and 30.0 cm^3 of 0.1 mol l^{-1} hydrochloric acid was added. Once the reaction was complete, the solution was made up to 250 cm^3 in a standard flask.
Describe the steps to prepare the 250 cm^3 solution. *(2 marks)*

© HERIOT-WATT UNIVERSITY

10.0 cm³ aliquots of the solution were titrated against 0.1 mol l⁻¹ standardised sodium hydroxide solution using phenolphthalein as an indicator until concordant results were obtained.

Q4: What do we mean by concordant results? *(1 mark)*
....

Q5: Why did the sodium hydroxide solution have to be standardised? *(1 mark)*

Benzocaine (see structure in the following diagram) is used to relieve pain and itching caused by conditions such as sunburn, insect bites or stings.

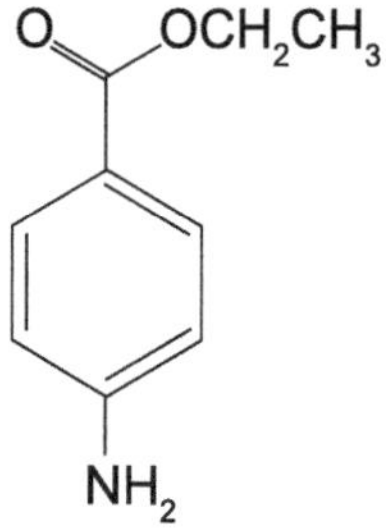

A student was carrying out a project to synthesise benzocaine. Part of the procedure to isolate the synthesised benzocaine is given as follows:

1. Add 20 cm³ of diethyl ether to the reaction mixture and pour into a separating funnel.
2. Add 20 cm³ of distilled water to the separating funnel.
3. Stopper the funnel, invert and gently shake.
4. Allow the aqueous layer to settle to the bottom.

Q6: Name the technique used. *(1 mark)*
....

Q7: Outline the next steps required to obtain a maximum yield of benzocaine. *(3 marks)*
....

Q8: State two properties the solvent must have to be appropriate. *(2 marks)*
....

Q9: Suggest a second technique that could be used to purify a solid sample of benzocaine. *(1 mark)*

TOPIC 6. PRACTICAL SKILLS AND TECHNIQUES

Thin layer chromatography (TLC) was used to help confirm the identity of the product. A small volume of solvent was used to dissolve a sample of product which was then spotted onto a TLC plate. The plate was allowed to develop (see following diagram).

Q10: Name the substance spotted at A on the TLC plate. *(1 mark)*
....

Q11: How pure is the student's product? *(1 mark)*

Unit 3 Topic 7
Researching chemistry test

Researching chemistry test

Q1: 2.58 g of hydrated barium chloride $BaCl_2.nH_2O$, was heated until constant mass was achieved, leaving 2.22 g of anhydrous barium chloride $BaCl_2$.

Calculate the value of *n* in the formula $BaCl_2.nH_2O$. Give your answer to the nearest whole number. *(2 marks)*

....

Q2: Oxalic acid, a dicarboxylic acid has the formula $H_2C_2O_4$ and can be made by reacting calcium oxalate with sulfuric acid.

$$H_2SO_4(aq) + CaC_2O_4(s) + xH_2O \rightarrow CaSO_4.xH_2O(s) + H_2C_2O_4(aq)$$

4.94 g of $CaSO_4.xH_2O$ was dehydrated to form 3.89 g of $CaSO_4$.

Calculate the value of *x* in the previous formula. Give your answer to the nearest whole number. *(2 marks)*

Q3: 10.0 cm^3 of a liquid drain cleaner containing sodium hydroxide was diluted to 250 cm^3 in a standard flask. 25.0 cm^3 samples of this diluted solution were pipetted into a conical flask and titrated against 0.220 mol l^{-1} sulfuric acid solution. The average of the concordant titres was 17.8 cm^3.

Calculate the mass of sodium hydroxide in 1 litre of drain cleaner. *(3 marks)*

Sulfa drugs are compounds with antibiotic properties. Sulfa drugs can be prepared from a solid compound called sulfanilamide which is prepared in a six stage synthesis. The equation for the final step in the synthesis is shown.

Q4: What type of reaction is this? *(1 mark)*

....

Q5: The sulfanilamide is separated from the reaction mixture and recrystallised from boiling water. Why is the recrystallisation necessary? *(1 mark)*

....

Q6: Calculate the percentage yield of sulfanilamide if 4.282 g of 4-acetamidobenzenesulfonamide produced 2.237 g of sulfanilamide. *(2 marks)*

....

Q7: Describe how a mixed melting point experiment would be carried out and the result used to confirm the product was pure. *(1 mark)*

....

Q8: Suggest another analytical technique which could be used to indicate whether the final sample is pure. *(1 mark)*

The formula of potassium hydrogen oxalate can be written as $K_xH_y(C_2O_4)_z$.

In an experiment to determine x, y and z, 4.49 g of this compound was dissolved in water and the solution made up to one litre. 20.0 cm^3 of the solution was pipetted into a conical flask and then titrated with 0.0200 mol l^{-1} acidified potassium permanganate at 60 °C. Average titre volume was 16.5 cm^3.

Equation for the reaction taking place is:

$$5C_2O_4^{2-} + 16H^+ + 2MnO_4^- \rightarrow 2Mn^{2+} + 10CO_2 + 8H_2O$$

Q9: What colour change would indicate the end of the titration? *(1 mark)*

....

Q10: Calculate the number of moles of oxalate ions, $C_2O_4^{2-}$, in 20.0 cm^3 of the solution. *(2 marks)*

....

Q11: Calculate the mass of oxalate ions in 1 litre of the solution. *(1 mark)*

....

Q12: Using another analytical procedure, 4.49 g of the potassium hydrogen oxalate was found to contain 0.060 g of hydrogen. Use this information from this and part c to calculate the mass of potassium in this sample. *(1 mark)*

....

Q13: Calculate the value of x, y, and z. *(2 marks)*

Organic Chemistry and Instrumental Analysis

1 Molecular orbitals	**245**
1.1 Formation of molecular bonding orbitals	246
1.2 Hybridisation and the role in formation of σ and π bonds	247
1.3 The bonding continuum	251
1.4 Summary	252
1.5 Resources	252
1.6 End of topic test	253
2 Synthesis	**255**
2.1 Practising with molecular structure	257
2.2 Homo- and heterolytic bond fission	260
2.3 Electrophiles and nucleophiles	262
2.4 Curly arrow notation	262
2.5 Haloalkanes	262
2.6 Alcohols	268
2.7 Ethers	273
2.8 Alkenes	274
2.9 Electrophilic addition	274
2.10 Carboxylic acids	279
2.11 Amines	282
2.12 Aromatic compounds	284
2.13 Reactions of benzene	286
2.14 Extra practice	292
2.15 Summary	295
2.16 Resources	295
2.17 End of topic test	296
3 Stereochemistry	**305**
3.1 Isomers	306
3.2 Geometric isomers	307
3.3 Optical isomers	310
3.4 Biological systems	312

3.5	Optical isomerism in medicines and other substances	314
3.6	Summary	315
3.7	Resources	315
3.8	End of topic test	316

4 Experimental determination of structure 323

4.1	Elemental microanalysis	324
4.2	Mass spectrometry	325
4.3	Infrared spectroscopy	332
4.4	Proton nuclear magnetic resonance (NMR) spectroscopy	337
4.5	Absorption of visible light	347
4.6	Chromophores	348
4.7	Summary	350
4.8	Resources	350
4.9	End of topic test	351

5 Pharmaceutical chemistry 357

5.1	Effect of drugs on the body	358
5.2	How drugs work	361
5.3	Pharmacological activity	361
5.4	Summary	362
5.5	Resources	362
5.6	End of topic test	363

6 Organic chemistry and instrumental analysis test 367

Unit 4 Topic 1

Molecular orbitals

Contents

- 1.1 Formation of molecular bonding orbitals 246
- 1.2 Hybridisation and the role in formation of σ and π bonds 247
- 1.3 The bonding continuum 251
- 1.4 Summary 252
- 1.5 Resources 252
- 1.6 End of topic test 253

Prerequisites

Before you begin this topic, you should know:

- what an orbital is and about the different types of orbital (s,p,d and f); (Advanced Higher Chemistry)
- the different types of bonding that occur within compounds and elements and how they fit into the bonding continuum. (National 5, Higher and Advanced Higher Chemistry)

Learning objective

By the end of this topic, you should know about:

- how molecular bonding orbitals are formed;
- how sigma and pi bonds are formed through hybridisation;
- the bonding continuum.

1.1 Formation of molecular bonding orbitals

When atoms approach each other, their separate sets of atomic orbitals merge to form a single set of molecular orbitals. Some of the molecular orbitals, known as 'bonding molecular orbitals', occupy the region between two nuclei. The attraction of positive nuclei to negative electrons occupying bonding molecular orbitals is the basis of bonding between atoms. Each molecular orbital can accommodate a maximum of two electrons.

If we look at the example of $H_2(g)$ one of the molecular orbitals is formed by adding the mathematical functions for the two 1s orbitals that come together to form this molecule. A molecular orbital is a mathematical function describing the wave-like behaviour of an electron in a molecule. This function can be used to calculate chemical and physical properties such as the probability of finding an electron in any specific region. This can be described simply as the region of space in which the function has a significant amplitude. Molecular orbitals are usually constructed by combining atomic orbitals or hybrid orbitals from each atom of the molecule, or other molecular orbitals from groups of atoms. The shape of orbitals are determined through quantum mechanics.

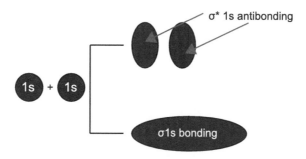

S-orbitals have a spherical symmetry surrounding a single nucleus, whereas σ-orbitals have a cylindrical symmetry and encompass two nuclei.

The bonding orbital is where the electrons spend most of their time between the two nuclei. Electrons in the other antibonding orbital spend most of their time away from the region between the two nuclei.

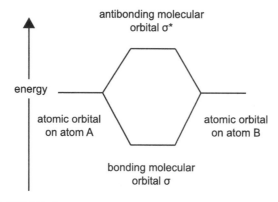

Electrons are added to orbitals in order of increasing energy. The two electrons associated with H_2 bonding are placed in the bonding molecular orbital suggesting that the energy of an H_2 molecule is lower than 2 separate H atoms. This results in a molecule of H_2 being more stable than two separate H atoms.

We can use this to explain why molecules of helium do not exist. Combining two helium atoms with the $1s^2$ electronic configuration would result in two electrons in the bonding orbital and two electrons in the antibonding orbital. The total energy of a He molecule would be exactly the same as two isolated He atoms with nothing to hold the atoms together in a molecule. For further atoms it is only the valence shell electrons used in combining atoms that will be considered as core electrons make no contribution to the stability of molecules.

1.2 Hybridisation and the role in formation of σ and π bonds

It is difficult to explain the shapes of even the simplest molecules with atomic orbitals. A solution to this problem was proposed by Linus Pauling, who argued that the valence orbitals on an atom could be combined to form hybrid atomic orbitals.

A covalent bond is formed when two half-filled atomic orbitals overlap. If they overlap along the axis of the bond (end on) a covalent bond is known as a sigma (σ) bond.

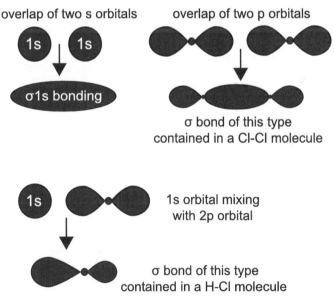

Pi (π) bonds arise where atoms make multiple bonds, for example the double bond in a molecule of oxygen O_2 is made up of one sigma and one pi bond. The triple bond in a molecule of nitrogen N_2 is made up of one sigma and two pi bonds. Pi bonds are formed when atomic orbitals lie perpendicular to the bond and overlap side on. End to end overlap is more efficient than side on overlap and therefore σ bonds are stronger than π bonds.

side-on overlap of two p orbitals in a π bond

If we look at carbon in its ground state it has the electronic configuration $1s^2\ 2s^2\ 2p^2$. This means it has two half-filled orbitals in the 2p subshell which may lead us to believe it would form two bonds instead of the four we know it forms. A simple explanation might involve promotion of an electron from the 2s orbital to the empty 2p orbital, producing four unpaired electrons which could then form four bonds with hydrogen.

A more satisfactory explanation involves hybridisation. The theory assumes that the 2s and three 2p orbitals combine during bonding to form four new identical hybrid orbitals.

The hybrid orbitals are known as sp^3 orbitals because they are formed by combining one s and three p orbitals.

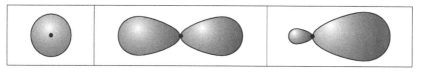

The energy required to promote the electron would be more than offset by the formation of two extra covalent bonds. However, whereas the others would involve 2p orbitals. Spectroscopic measurements show that all four bonds in methane are identical.

Let's look at an alkane, ethane for example. Each carbon has three 2p orbitals and one 2s orbital which mix to form four degenerate (equal energy) hybrid orbitals. These are known as sp^3 hybrid orbitals and point towards the corners of a tetrahedron in order to minimise repulsion from each other. The four sp^3 orbitals on each carbon atom overlap end to end with one sp^3 orbital on the other carbon atom and the three hydrogen 1s orbitals. This forms 4 σ bonds.

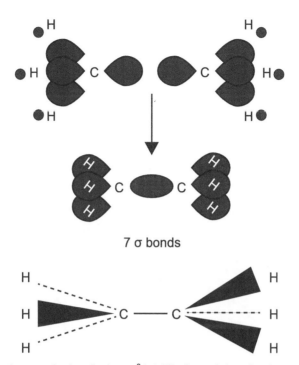

7 σ bonds

The bonding in ethane can be described as sp^3 hybridisation and sigma bonds.

If we take a look at the bonding in the corresponding alkene ethene we can see that the 2s orbital and two of the three 2p orbitals mix on each carbon atom to form three sp^2 hybrid orbitals. To minimise repulsion these orbitals form a trigonal planar arrangement. The carbon atoms use the three sp^2 hybrid orbitals to form sigma bonds with two hydrogen atoms and with the other carbon atom. The unhybridised 2p orbitals left on the carbon can overlap side-on to form a pi bond.

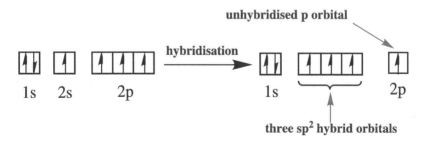

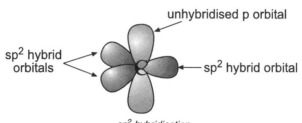

sp² hybridisation

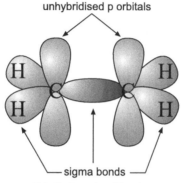

C to C sigma bond forming

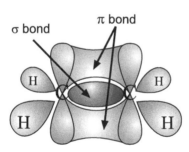

In alkynes for example ethyne (C_2H_2) the 2s orbital and one of the three 2p orbitals mix on each carbon to form two sp hybrid orbitals. This will form 3 sigma bonds and 2 pi bonds. The alkyne adopts a linear structure to minimise repulsion.

TOPIC 1. MOLECULAR ORBITALS

> **Bonding in hydrocarbons**   Go online
>
> It may help to draw Lewis dot structures for the molecule to identify the bonding and non-bonding pairs of electrons before answering the question.
>
> For the next three questions, consider the hydrocarbon ethyne, HC≡CH.
>
> **Q1:** What is the hybridisation of the carbon atoms required to form a C≡C bond?
>
> a) sp^3
> b) sp^2
> c) sp
>
> ....
>
> **Q2:** Which of the following statements is true about an ethyne molecule?
>
> a) There are three π bonds and two σ bonds.
> b) There are two π bonds and two σ bonds.
> c) There are two π bonds and three σ bonds.
> d) There are three π bonds and three σ bonds.
>
> ....
>
> **Q3:** What shape will the ethyne molecule be?

1.3 The bonding continuum

The symmetry and position of the bonding orbitals between atoms will determine the type of bonding: ionic, polar covalent or pure covalent.

For pure covalent bonds where there is no or very little difference in electronegativity between the atoms involved in the bond the bonding orbital will be symmetrical around a point where the bonding electrons are predicted to be found. A fluorine molecule F_2 would be an example of where this would be found.

In polar covalent bonds ($H^{\delta+}$-$Cl^{\delta-}$) where there is a difference in electronegativity between the two atoms involved in the bond the bonding orbital will be asymmetrical with the bonding electrons more likely to found around chlorine due to it having a larger electronegativity.

In ionic bonds (Na^+Cl^-) where the difference in electronegativity between the two atoms involved in the bond is large, the bonding orbital will be extremely asymmetrical and almost entirely around one of the atoms. In this case it would be around chlorine due to it having a larger electronegativity compared to sodium.

1.4 Summary

> **Summary**
>
> You should now be able to state that:
>
> - molecular bonding orbitals are formed by the overlap of orbitals from individual atoms when bonding.
> - hybridisation occurs when different types of atomic orbitals are mixed.
> - sigma bonds are formed when two half filled atomic orbitals overlap end-on.
> - pi bonds are formed when two half filled atomic orbitals overlap side-on.
> - where compounds fit onto the bonding continuum is determined by the difference in electronegativity between bonded atoms.

1.5 Resources

- LearnChemistry, Chemistry Vignettes: Molecular Orbitals (http://rsc.li/2a1zkg0)
- Molecular Orbital Theory (http://bit.ly/2abKNqV)

1.6 End of topic test

End of Topic 1 test   Go online

Q4: Which is the correct description of the numbers of sigma and pi bonds in propene?

a) 1 sigma bond, 8 pi bonds
b) 2 sigma bonds, 8 pi bonds
c) 7 sigma bonds, 1 pi bonds
d) 8 sigma bonds, 1 pi bond

....

Q5: The bonds in 2-bromobutane ($CH_3CH_2CHBrCH_3$) are sp^3 hybridised. Explain what is meant by sp^3 hybridisation.

....

Q6: How many sp^2 hybridised carbon atoms does the following molecule contain?

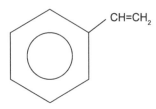

a) 0
b) 2
c) 6
d) 8

....

Q7: Ethene molecules contain:

a) sp^2 hybridised carbon atoms but no sp^3 hybridised carbon atoms.
b) sigma bonds but no pi bonds.
c) sp^3 hybridised carbon atoms but no sp^2 hybridised carbon atoms.
d) pi bonds but no sigma bonds.

....

Q8: Predict the bonding in $TiCl_4$ and where the bonding orbitals will be found.

Unit 4 Topic 2

Synthesis

Contents

2.1	Practising with molecular structure	257
2.2	Homo- and heterolytic bond fission	260
2.3	Electrophiles and nucleophiles	262
2.4	Curly arrow notation	262
2.5	Haloalkanes	262
2.6	Alcohols	268
2.7	Ethers	273
2.8	Alkenes	274
2.9	Electrophilic addition	274
2.10	Carboxylic acids	279
2.11	Amines	282
2.12	Aromatic compounds	284
2.13	Reactions of benzene	286
2.14	Extra practice	292
2.15	Summary	295
2.16	Resources	295
2.17	End of topic test	296

Prerequisites

Before you begin this topic, you should know how to:

- determine the molecular formula of a compound. (National 5 Chemistry)
- draw the structural formula of a compound. (National 5 Chemistry)

256 UNIT 4. ORGANIC CHEMISTRY AND INSTRUMENTAL ANALYSIS

Learning objective

By the end of this topic, you should be able to:

- how to draw the skeletal structure of a compound from the molecular formula or the structural formula and vice versa.
- recognise and use different types of reaction in organic synthesis including substitution, addition, elimination, condensation, hydrolysis, oxidation, reduction;
- devise synthetic routes, with no more than three steps, from a given reactant to a final product;
- look at molecular structures and deduce the reactions they can undergo.

2.1 Practising with molecular structure

There are many representations of organic molecules and you need to be able to convert easily between the different forms. These include molecular formulae and structural formulae of which you should be familiar with but we can also represent molecules in skeletal formulae.

The molecular formula tells us the number of each type of atom within an organic molecule. For example the molecular formula of butane is C_4H_{10} telling us we have a molecule with four carbon atoms and ten hydrogen atoms. The following table shows some molecular formulae for a few selected hydrocarbon molecules.

Hydrocarbon molecule	Molecular formula
Ethane	C_2H_6
Propane	C_3H_8
Butene	C_4H_8
Pentene	C_5H_{10}
Cyclobutane	C_4H_8
Cyclohexane	C_6H_{12}
Ethyne	C_2H_2

For molecules with functional groups including hydroxyl (OH), amine (NH_2), carboxyl (COOH) and carbonyl (CO) these are shown as part of the molecular formula of the molecule. Some examples are given in the following table.

Organic molecule	Molecular formula
Ethanol	C_2H_5OH
Ethanoic acid	CH_3COOH
Ethylamine	$C_2H_5NH_2$
Ethanal	CH_3CHO

If we go back to the example of butane we can represent this molecule through its structural formula. This gives a representation of how the elements are bonded within the molecule.

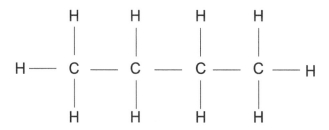

This shows us that the four carbons are singly bonded in a straight chain with the appropriate number of hydrogen atoms bonded to each carbon.

The structure of butane can also be represented as a skeletal formula. In organic chemistry, skeletal formulae are the most abbreviated diagrammatic descriptions of molecules in common use. They look very bare because in skeletal formulae the hydrogen atoms (attached directly to carbons) are removed, leaving just a 'carbon skeleton' with functional groups attached to it. You must remember that hydrogen atoms that are not shown in skeletal formula are assumed to be there.

Some points to note with skeletal formulae:

a) There is a carbon atom at each junction between bonds in a chain and at the end of a bond (unless there is something else there like a functional group);

b) There are enough hydrogen atoms attached to each carbon atom to make the total number of bonds of each carbon four.

So looking at butane again, the skeletal formula would be:

The four carbon atoms are positioned in the following places in the skeletal formula. Attached to these carbons would be the appropriate number of hydrogen atoms to give each carbon four bonds.

If we look at the molecule 2-methylbutane the skeletal structure would look like this:

    skeletal formula   structural formula

If we look at the skeletal formula of ethanol, for example, the functional group OH is included in the formula.

the two carbon atoms are positioned here with the appropriate number of hydrogen atoms attached

$$H-\underset{\underset{H}{|}}{\overset{\overset{H}{|}}{C}}-\underset{\underset{H}{|}}{\overset{\overset{H}{|}}{C}}-O-H$$

Some other molecules drawn as skeletal formulae are given as follows.

Amino acid asparagine

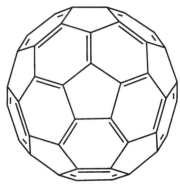

Buckminsterfullerene C60 (http://commons.wikimedia.org/wiki/File:C60a.png, by http://en.wikipedia.org/wiki/User:Mstroeck, **licensed under** http://creativecommons.org/licenses/by-sa/3.0/deed.en)

Cholesterol

You should practise converting between the molecular, structural and skeletal formulae for various molecules in particular the different families of hydrocarbons studied at National 5, Higher and Advanced Higher level. This should be done for molecules with up to 10 carbon atoms in the longest chain.

2.2 Homo- and heterolytic bond fission

Alkanes are not particularly reactive due to the non-polar nature of their bonds. They can however, react with halogens in the presence of sunlight or UV light where halogenoalkanes are produced along with steamy fumes of the corresponding hydrogen halide. In this reaction an atom of hydrogen has been replaced with an atom of a halogen and is an example of a substitution reaction. This substitution reaction is thought to occur by a chain reaction which has three main steps: initiation, propagation and termination (chain reactions have already been covered briefly in the CfE Higher

TOPIC 2. SYNTHESIS

course and you will be familiar with these steps).

Bond breaking: if we look at the reaction between methane and chlorine:

$$CH_4 + Cl_2 \rightarrow CH_3Cl + HCl$$

This reaction does not take place in the dark, it requires UV light to provide the energy to break the chlorine-chlorine bond. This splits the chlorine molecules into chlorine atoms.

$$Cl - Cl \rightarrow Cl\cdot + Cl\cdot \text{ (the dot represents an unpaired electron)}$$

This type of bond breaking is known as homolytic fission and usually occurs when the bond is non polar or very slightly polar. In homolytic fission one electron from the bond goes to one atom while the other electron goes to the other atom. Atoms with unpaired electrons are known as radicals which are incredibly unstable and therefore extremely reactive. The initiation step in a chain reaction produces radicals.

Note: If a bond were to split unevenly (one atom getting both electrons, and the other none), ions would be formed. The atom that got both electrons would become negatively charged, while the other one would become positive. This is called heterolytic fission and will be favoured when the bond is polar. Reactions proceeding via heterolytic fission tend to produce far fewer products and are therefore better suited for synthesis.

$$A - B \rightarrow A^{\oplus} + B^{\ominus}$$
$$A - B \rightarrow A^{\ominus} + B^{\oplus}$$

Heterolytic fission

Initiation steps are followed by propagation and termination steps in chain reactions. Looking at the reaction between methane and bromine the propagation and termination steps can be seen as follows.

Chain initiation

The chain is initiated by UV light breaking some bromine molecules into free radicals by homolytic fission.

$$Br_2 \rightarrow 2Br\cdot$$

Chain propagation reactions

These are the reactions which keep the chain going. In each of these propagation steps one radical enters the reaction and another is formed.

$$CH_4 + Br\cdot \rightarrow CH_3\cdot + HBr$$
$$CH_3\cdot + Br_2 \rightarrow CH_3Br + Br\cdot$$

Chain termination reactions

These are reactions which remove free radicals from the system without replacing them by new ones. This brings the chain reaction to an end.

$$Br\cdot + Br\cdot \rightarrow Br_2$$
$$CH_3\cdot + Br\cdot \rightarrow CH_3Br$$
$$CH_3\cdot + CH_3\cdot \rightarrow CH_3CH_3$$

2.3 Electrophiles and nucleophiles

In reactions involving heterolytic bond fission, attacking groups are classified as 'nucleophiles' or 'electrophiles'.

Electrophiles are chemical species that are electron deficient and therefore are 'electron loving' species. Electrophiles are molecules or positively charged ions which are capable of accepting an electron pair. They will seek out electron rich sites in organic molecules; examples include NO_2^+ and SO_3H^+.

Nucleophiles are chemical species that are electron rich and are 'electron donating' species. Nucleophiles are molecules or negatively charged ions which have at least one lone pair of electrons that they can donate and form dative bonds. They will seek out electron-deficient sites in organic molecules; examples include H_2O, NH_3 and the halide ions.

2.4 Curly arrow notation

Double headed curly arrows are used to indicate the movement of electron pairs in a reaction. The tail of the arrow shows where the electrons originate from and the head shows where they end up. An arrow starting at the middle of a covalent bond indicates that heterolytic bond fission is occurring. When an arrow is drawn with the head pointing to the space between two atoms, this indicates that a covalent bond will be formed between the two atoms.

A single headed curly arrow indicates the movement of a single electron. These are useful in discussions about radical chemistry mechanisms.

2.5 Haloalkanes

Haloalkanes can be regarded as substituted alkanes where one or more of the hydrogen atoms have been replaced with a halogen atom. In naming of haloalkanes the halogen atoms are treated as branches and naming is done in the same way as for branched alkanes.

```
        Br    F
        |    |
Cl ─── C ─── C ─── F
        |    |
        H    F
```

This haloalkane is named 2-bromo-2-chloro-1,1,1-trifluoroethane (remember branches are named in alphabetical order)

TOPIC 2. SYNTHESIS

Monohaloalkanes (where one H atom is substituted for a halogen atom) have three different structural types which are primary, secondary and tertiary. These are determined by the number of alkyl groups attached to the carbon atom that is directly attached to the halogen atom. In these diagrams, X represents a halogen atom:

Primary monohaloalkane, one alkyl group attached to the carbon atom directly attached to the halogen atom

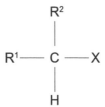

Secondary monohaloalkane, two alkyl groups attached to the carbon atom directly attached to the halogen atom

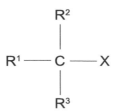

Tertiary monohaloalkane, three alkyl groups attached to the carbon atom directly attached to the halogen atom

Due to the polar nature of the carbon-halogen bond haloalkanes are susceptible to nucleophilic attack.

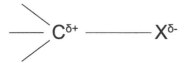

The presence of the slight positive charge on the carbon atom makes haloalkanes susceptible to nucleophilic attack. The nucleophile donates a pair of electrons forming a bond with the carbon atom of the C-X bond. The halogen atom is 'thrown out' and substituted by the nucleophile. The mechanism for this is included in the Haloalkanes section of this topic under S_N1 and S_N2 reactions.

© HERIOT-WATT UNIVERSITY

Nucleophilic substitutions

Reaction of monohaloalkanes with alkalis produces alcohols. A solution of aqueous potassium hydroxide (KOH) or sodium hydroxide (NaOH) is used.

$$CH_3-CH_2-CH_2-Cl \xrightarrow[\text{Aqueous KOH/NaOH}]{OH^-} CH_3-CH_2-CH_2-OH$$

In this reaction the nucleophile is OH^-

Reaction with alcoholic potassium alkoxide (potassium methoxide in methanol CH_3OK) produces ethers.

$$CH_3-CH_2-CH_2-Cl \xrightarrow[\text{Potasssium methoxide in methanol}]{CH_3O^-} CH_3-CH_2-CH_2-O-CH_3$$

Methoxypropane (ether)

In this reaction the nucleophile is CH_3O^-

Reaction with ethanolic potassium cyanide or sodium cyanide (KCN or NaCN in ethanol) produces nitriles.

$$CH_3-CH_2-CH_2-Cl \xrightarrow[\text{Ethanolic KCN or NaCN}]{CN^-} CH_3-CH_2-CH_2-CN$$

Butanenitrile (nitrile)

The nucleophile in this reaction is CN^-

The end nitrile contains one more carbon than the original haloalkane. This is very useful in synthetic organic chemistry as a way of increasing the chain length of an organic compound. The nitrile can be converted into the corresponding carboxylic acid through acid hydrolysis.

$$CH_3-CH_2-CH_2-CN \xrightarrow{H_2O/H^+} CH_3-CH_2-CH_2-COOH$$

Monohaloalkanes can also undergo elimination reactions to form alkenes. This is achieved by heating the monohaloalkane under reflux with ethanolic potassium or sodium hydroxide.

TOPIC 2. SYNTHESIS

$$\underset{\text{2-bromopropane}}{\text{H}_3\text{C}-\text{CHBr}-\text{CH}_3} \xrightarrow{\text{OH}^-} \underset{\text{propene}}{\text{H}_2\text{C}=\text{CH}-\text{CH}_3} + \text{Br}^- + \text{H}_2\text{O}$$

In this reaction a hydrogen halide is removed from the original monohaloalkane and for some it can result in two different alkenes being produced. This is due to the availability of more than one H atom that can be removed in the formation of the hydrogen halide. For example 2-chlorobutane can result in but-1-ene and but-2-ene, of which but-2-ene is the major product.

The resulting alkene can be tested by the addition of bromine water which will be decolourised in the presence of an alkene due to the bromine being added across the double bond in an addition reaction.

The presence of halogen ions can be tested by reacting the substance with silver nitrate solution. The silver halide is precipitated due to them being insoluble in water. The colour of the precipitate can tell you which halide ion was originally present in the substance.

Silver halide	Colour of precipitate
Sliver chloride	White
Silver bromide	Cream
Silver Iodide	Yellow

It is quite difficult to tell the colours of these precipitates apart especially if very little precipitate is produced. You can add the precipitate to ammonia solution and confirm which halide is present.

Original precipitate	Observation
AgCl	Precipitate dissolves to give a colourless solution.
AgBr	Precipitate is almost unchanged using dilute ammonia solution, but dissolves in concentrated ammonia solution to give a colourless solution.
AgI	Precipitate is insoluble in ammonia solution of any concentration.

Haloalkanes have been used as anticancer agents however are known for their toxic side effects. They are also often used as alkylating agents.

S_N1 and S_N2 reactions

Haloalkanes will undergo **nucleophilic substitution** by one of two different reaction mechanisms which are called S_N1 and S_N2.

S_N1 reaction mechanism

A kinetic study of the reaction between 2-bromo-methylpropane (tertiary haloalkane) and the nucleophile OH^- shows it has the rate equation rate=$k[(CH_3)_3CBr]$. This means it is first order with respect to the haloalkane implying the rate determining step can only involve the haloalkane. It is a two-step process. The polarity of the C-Br bond is shown in Step 1.

Step 1

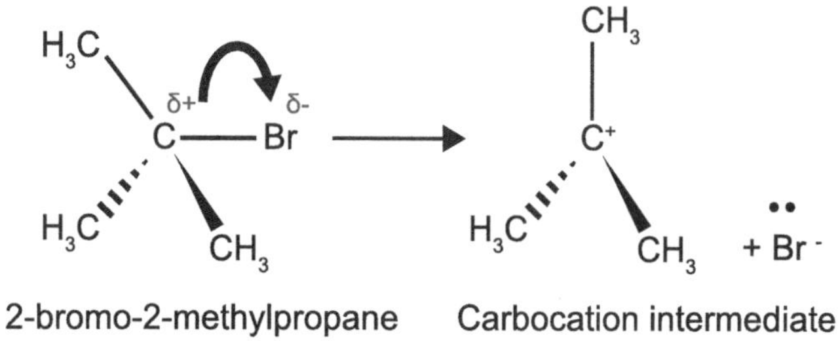

2-bromo-2-methylpropane   Carbocation intermediate

The first step produces a carbocation intermediate

Step 2

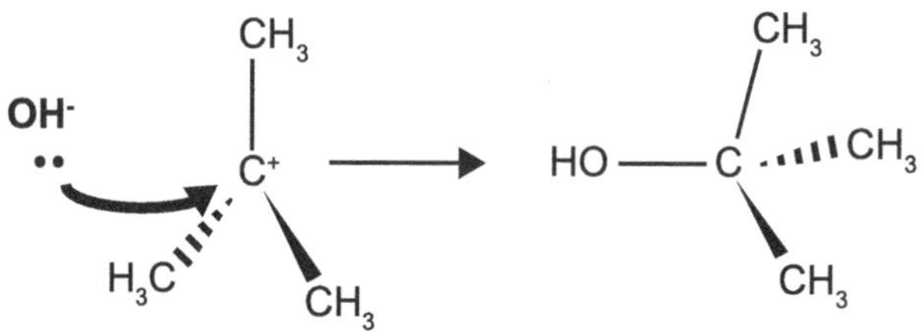

S_N2 reaction mechanism

A kinetic study of the reaction of bromoethane (primary haloalkane) and the nucleophile OH^- has the rate equation rate=$k[CH_3CH_2Br][OH^-]$. This means it is first order with respect to both the haloalkane and the hydroxide ion implying the rate determining step should involve both these species. This is a **one-step process**.

Step 1

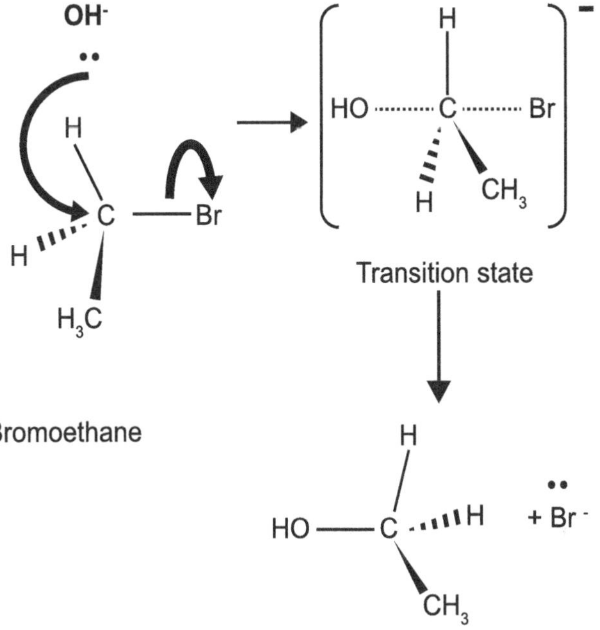

[Diagram: OH⁻ attacks bromoethane; transition state [HO·····C·····Br]⁻ with H, H, CH₃; produces HO-C(H)(H)(CH₃) + Br⁻]

Bromoethane

Transition state

The negative hydroxide ion in the bromoethane is a nucleophile so attacks the slightly positive carbon atom. A negatively charged intermediate is formed.

How do you know which reaction mechanism (S_N1 or S_N2) a haloalkane will undergo? You need to look at what type of haloalkane primary, secondary or tertiary you are dealing with. In the S_N1 reaction a **carbocation** intermediate is formed which could be a primary, secondary or tertiary carbocation but since alkyl groups are electron donating the tertiary carbocation will be the most stable. Tertiary haloalkanes are the most likely and primary haloalkanes the least likely to proceed by a S_N1 reaction mechanism.

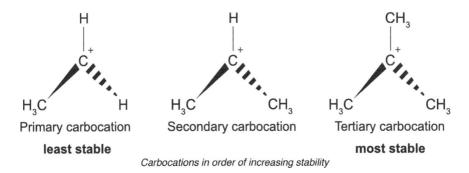

Primary carbocation — **least stable**
Secondary carbocation
Tertiary carbocation — **most stable**

Carbocations in order of increasing stability

In S_N1 reactions a positively charged intermediate is formed, whereas in S_N2 reactions a negatively charged intermediate is formed.

In a S_N2 reaction the OH⁻ nucleophile attacks the carbon atom of the carbon-halogen bond from the side opposite to the halogen atom. In the case of tertiary haloalkanes that position is most likely to be hindered by three bulky alkyl groups. Tertiary haloalkanes are least likely and primary haloalkanes most likely to proceed by a S_N2 reaction mechanism.

2.6 Alcohols

Prior knowledge from National 5 and Higher Chemistry is required for this section on alcohols. This includes:

a) Basic structure of alcohols including the functional group hydroxyl (OH);

b) Structural types of alcohols including primary, secondary and tertiary;

c) Oxidation reactions of alcohols.

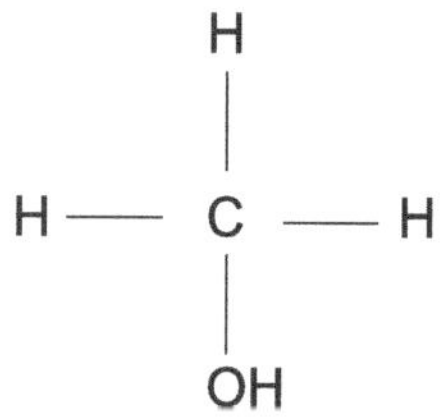

Methanol: first member of the alcohol family containing the hydroxyl (OH) functional group

Properties of alcohols

As the chain length of alcohols increase with the addition of a CH_2 unit between each progressive member their boiling points show a progressive increase. However, if we compare the boiling point of an alcohol to the boiling point of an alkane of similar relative formula mass and shape we can see that they are considerably higher. This is due to the presence of the polar hydroxyl (OH) group in the alcohol molecule allowing hydrogen bonding to be set up between the individual molecules. This is shown in the following diagram.

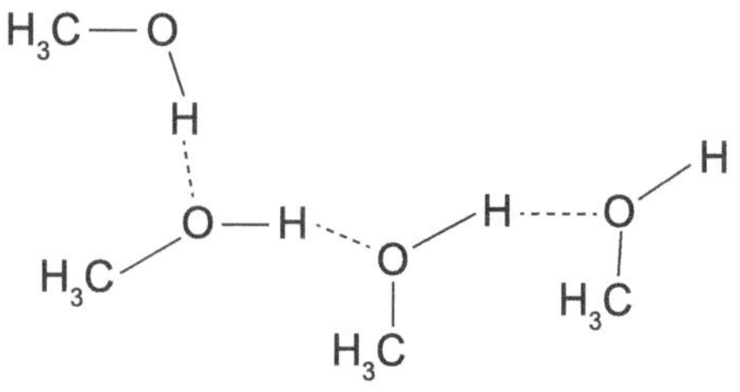

Hydrogen bonding between methanol molecules (shown as dashed lines)

TOPIC 2. SYNTHESIS

Between the alkane molecule of similar relative formula mass and shape only London dispersion forces are found and since hydrogen bonds are stronger extra energy is required to break them giving a reason for the higher boiling point of alcohols.

$$H-\underset{\underset{H}{|}}{\overset{\overset{H}{|}}{C}}-\underset{\underset{H}{|}}{\overset{\overset{H}{|}}{C}}-H$$

Ethane is the alkane closest to methanol in terms of relative formula mass (methanol 32 g and ethane 30 g). Only London dispersion forces are found between individual molecules of ethane. The boiling point of methanol is 64.7° C and the boiling point of ethane is -89° C showing the increased effect that hydrogen bonding has on the boiling point of methanol.

There is also a graduated decrease in the solubilites of alcohols in water as the chain length of the alcohol increases. Lower chain length alcohols (methanol, ethanol and propan-1-ol) are completely soluble in water (miscible with water) but alcohols such as heptan-1-ol and other higher chain length alcohols are insoluble in water. The smaller chain alcohols are soluble in water as the energy released in forming hydrogen bonds between the alcohol and water molecules is enough to break the hydrogen bonds between the water molecules. By the time you reach heptan-1-ol the large non-polar hydrocarbon part of the molecule disrupts the hydrogen bonding ability of the water with the hydroxyl, hence reducing solubility in water.

Preparation of alcohols

Alcohols can be prepared by two different reactions:

a) Heating haloalkanes under reflux with aqueous sodium/potassium hydroxide by nucleophilic substitution (see the Haloalkane section of this topic under nucleophilic substitution reactions);

b) Acid catalysed hydration of alkenes described as follows.

Alkenes undergo addition reactions with water to form alcohols. This reaction is an acid catalysed hydration proceeding through a carbocation intermediate.

Step 1

The hydrogen ion of the acid catalyst is an electrophile and the electrons of the double bond in the alkene (electron rich) attack the hydrogen ion forming a carbocation.

© HERIOT-WATT UNIVERSITY

See Markovnikov's rule, discussed in the 'Electrophilic addition' section later in this topic, to find out why the hydrogen is more likely to attach to that particular carbon in step 1.

Step 2

The carbocation undergoes rapid nucleophilic attack by a water molecule to give a protonated alcohol (alcohol with a hydrogen ion attached).

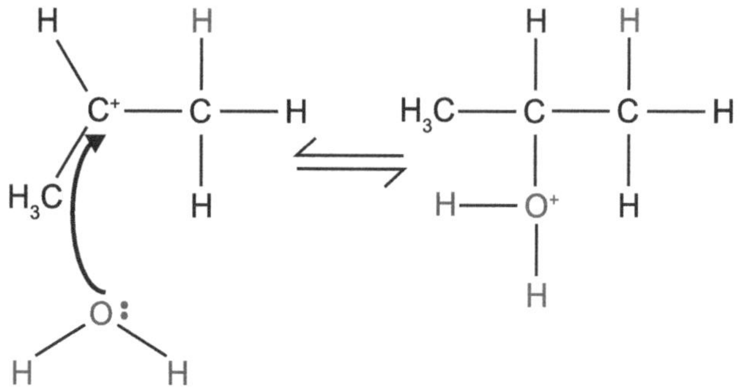

Step 3

Formation of the alcohol propan-2-ol. The protonated propan-2-ol is a strong acid and readily loses a proton to give propan-2-ol the final product.

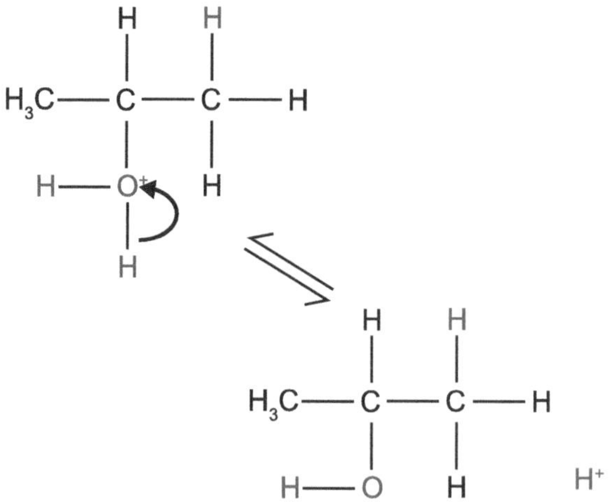

Alkoxides

Mentioned briefly in the Haloalkanes section, alkoxides are formed by adding an alkali metal such as potassium or sodium with an alcohol. For example when potassium is added to methanol, potassium methoxide is formed.

$$2K + 2CH_3OH \rightarrow H_2 + 2CH_3O^-K^+$$

Dehydration of alcohols

Dehydration of alcohols forms alkenes. This can be done in two ways either by passing the vapour of the alcohol over hot aluminium oxide or by treating the alcohol with concentrated sulfuric acid or phosphoric acid (orthophosphoric acid).

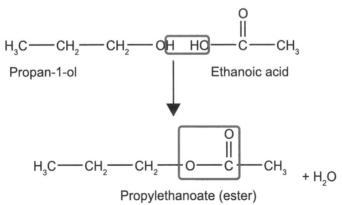

During dehydration the **OH** is removed along with an **H** atom on an adjacent carbon. This forms 2 alkenes with but-2-ene being the major product. With some alcohols such as propan-2-ol and butan-1-ol only one alkene is produced.

Esters

Alcohols can be reacted with carboxylic acids or acid chlorides to form esters. This is a condensation or esterification reaction carried out with a catalyst of concentrated sulfuric acid if using the carboxylic acid. When using an acid chloride the reaction is much faster and a catalyst is not required. Esters have been covered at National 5 and Higher level.

Ester linkage formed from the loss of water between the alcohol and carboxylic acid

$H_3C—CH_2—CH_2—O\boxed{H \quad Cl}—\underset{\underset{\displaystyle \|}{O}}{C}—CH_3$

Propan-1-ol — Ethanoyl chloride (acid chloride)

↓

$H_3C—CH_2—CH_2—\boxed{O—\underset{\underset{\displaystyle \|}{O}}{C}}—CH_3$ + HCl

Propylethanoate (ester)

Ester linkage formed from the loss of HCl between the alcohol and acid chloride

Reduction of aldehydes and ketones

Aldehydes formed from the mild oxidation of primary alcohols using hot copper (II) oxide or acidified potassium dichromate can be reduced back to primary alcohols by reacting with lithium aluminium hydride ($LiAlH_4$) dissolved in ether. Sodium borohydride ($NaBH_4$) can also be used as a reducing agent. Similarly ketones formed from the mild oxidation of secondary alcohols can be reduced back to secondary alcohols.

Aldehyde → (Reduction, $LiAlH_4$) → Primary alcohol

Ketone → (Reduction, $LiAlH_4$) → Secondary alcohol

The reactions are usually carried out in solution in a carefully dried ether such as ethoxyethane (diethyl ether).

Alcohol hydroxyl groups are present in a lot of pharmaceutical drugs as they are involved in hydrogen bonding with protein binding sites particularly beta blockers and anti-asthmatics.

2.7 Ethers

Ethers are synthesised from haloalkanes and alkoxides. They were the first anaesthetics and have the general formula R-O-R' where R and R' are alkyl groups. If R and R' are different then the ether is unsymmetrical and when identical they are symmetrical.

The reaction of 1-chloropropane with potassium methoxide in methanol produces methoxypropane (ether).

Ethers are named by assigning the longest carbon chain as the parent name. This is prefixed by the alkoxy substituent which has been named by removing the 'yl' from the name of the alkyl substituent and adding 'oxy'. CH_3CH_2O- is named ethoxy and CH_3O- is named methoxy, for example.

2-ethoxypropane: longest chain in this ether is propane; alkoxy substituent (attached to the 2nd carbon).

Ethoxypropane

Properties of ethers

The boiling points of ethers are much lower than that of their isomeric alcohols due to the fact that hydrogen bonding does not occur between ether molecules. This is due to the highly electronegative oxygen atom not being directly bonded to a hydrogen atom. They can however form hydrogen bonds with water molecules as shown.

Hydrogen bonding can occur between water and ether molecules but not between ether molecules

This also explains why low relative formula mass ethers are soluble in water for example methoxymethane and methoxyethane. The larger ethers are insoluble in water and therefore are useful in extracting organic compounds from aqueous solutions.

Ethers are highly flammable and when exposed to air slowly form peroxides which are unstable and can be explosive. Ethers are used as solvents as most organic compounds dissolve in them and they are relatively chemically inert. They are easily removed by distillation due to being volatile.

2.8 Alkenes

Alkenes can be prepared in the laboratory by:

a) dehydration of alcohols using aluminium oxide, concentrated sulfuric acid or orthophosphoric acid (see the 'Alcohols' section of this topic).

b) base-induced elimination of hydrogen halides from monohaloalkanes (see the 'Haloalkanes' section of this topic).

2.9 Electrophilic addition

Alkenes undergo **electrophilic addition** reactions with a variety of substances to form different products.

Catalytic addition of hydrogen

$$CH_3-CH=CH-CH_3 \quad \text{But-2-ene}$$

$$\xrightarrow{H-H}$$

$$CH_3-CH_2-CH_2-CH_3 \quad \text{Butane}$$

This reaction is also known as hydrogenation and is catalysed by nickel or palladium.

TOPIC 2. SYNTHESIS

Addition of halogens (halogenation)

The mechanism for the addition of halogens to alkenes involves two steps.

Step 1

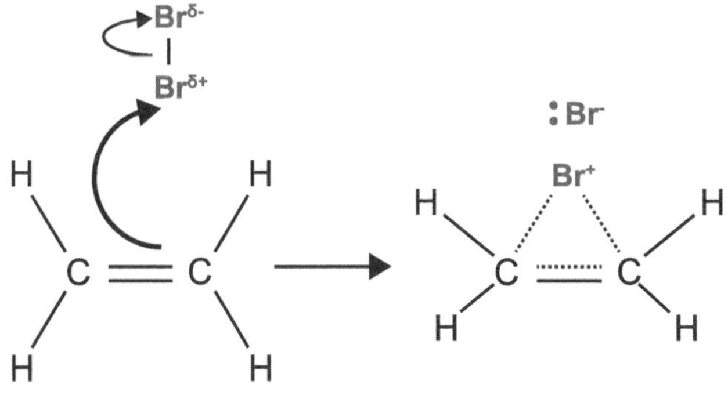

Ethene → Bromonium ion

The bromine molecule is the electrophile in this reaction and undergoes heterolytic fission. It approaches the double bond in ethene and becomes polarised. The electron rich double bond pushes the electrons in the bromine molecule towards the bromine atom which is furthest away from the double bond which gains a slight negative charge. The other bromine atom gains a slight positive charge and the Br-Br bond breaks heterolytically creating a cyclic intermediate and a bromide ion.

Step 2

The bromide ion attacks the cyclic intermediate ion from the opposite side to the Br atom which prevents access to the side where it is located. The bromide ion is acting as a nucleophile seeking out a centre of positive charge.

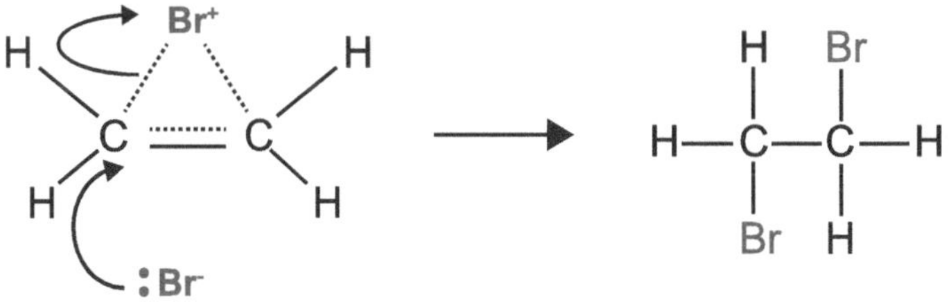

Cyclic ion intermediate → 1,2-dibromoethane

Addition of hydrogen halides (hydrohalogenation)

The mechanism for this addition is also a two-step process.

Step 1

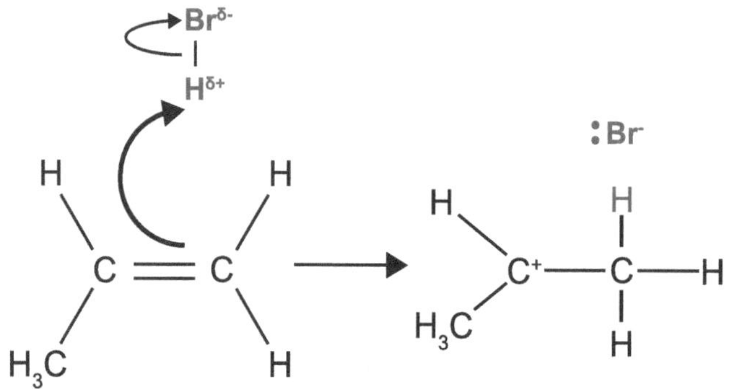

The H-Br molecule is already polarised and the electrons of the double bond attack the hydrogen. The H-Br bond breaks heterolytically and a bromide ion is formed at the same time. A carbocation is formed after the double bond breaks to form a new bond to the hydrogen.

Step 2

The second step involves the bromide ion attacking the carbocation intermediate which can be done from either side of the carbocation.

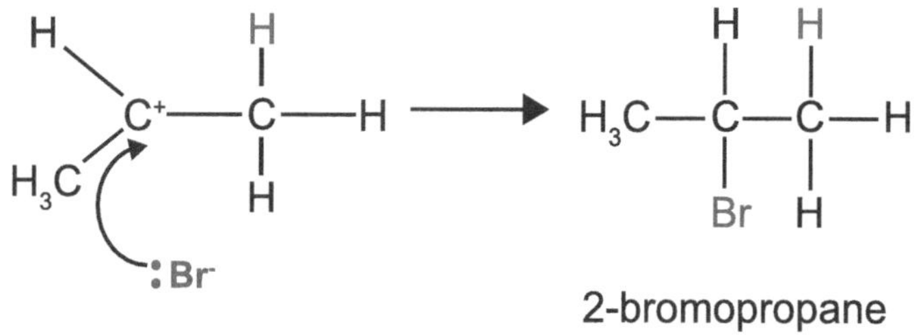

2-bromopropane

In the diagram the product is 2-bromopropane but as propane is an unsymmetrical alkane 1-bromopropane will also be formed.

Markovnikov's rule

When a hydrogen halide is added to an unsymmetrical alkene (one where the groups attached to one carbon of the double bond are different from the groups attached to the other carbon of the double bond) two products are formed. Markovnikov's rule states that when H-X is added onto an unsymmetrical alkene the major product is the one where the hydrogen bonds to the carbon atom of the double bond that has already the greatest number of hydrogen atoms attached to it.

TOPIC 2. SYNTHESIS

$$H-\overset{\overset{\displaystyle H}{|}}{\underset{\underset{\displaystyle H}{|}}{C}}-\overset{\overset{\displaystyle H}{|}}{\underset{\underset{\displaystyle H}{|}}{C}}-\overset{\displaystyle H}{\underset{\displaystyle H}{C}}=\overset{\displaystyle H}{\underset{\displaystyle H}{C}}-H$$

If we add for example H-Cl onto but-1-ene the hydrogen atom should attach itself to the carbon atom shown in blue as it has the greatest number of hydrogen atoms already attached to it. This would form the product 2-chlorobutane as the major product.

$$H-\overset{H}{\underset{H}{C}}-\overset{H}{\underset{H}{C}}-\overset{Cl}{\underset{H}{C}}-\overset{H}{\underset{H}{C}}-H \quad \text{2-chlorobutane}$$

The minor product would be 1-chlorobutane:

$$H-\overset{H}{\underset{H}{C}}-\overset{H}{\underset{H}{C}}-\overset{H}{\underset{H}{C}}-\overset{Cl}{\underset{H}{C}}-H \quad \text{1-chlorobutane}$$

Addition of water (hydration)

The addition of water which is catalysed by an acid has a very similar mechanism for the addition of hydrogen halides proceeding through a carbocation intermediate. This has already been covered in the Alcohols section of this topic but has been included again in this section.

Step 1: formation of a carbocation

Step 2: the carbocation undergoes nucleophilic attack by a water molecule to give a protonated alcohol

Step 3: the protonated alcohol which is a strong acid loses a proton to give the final alcohol

This reaction also follows Markovnikov's rule in determining the major product when water is added to an unsymmetrical alkene. The major product in this reaction is propan-2-ol with propan-1-ol formed as the minor product also. This is also due to the stability of the carbocation formed in the first step.

2.10 Carboxylic acids

Preparation of carboxylic acids

Carboxylic acids can be prepared by:

a) oxidising primary alcohols or aldehydes by heating them with acidified potassium dichromate (this has been covered at higher chemistry and should be revised).

b) hydrolysis of nitriles, esters or amides by heating them in the presence of a catalyst which can be either an acid or an alkali. The hydrolysis of nitriles was covered earlier in the section 'Haloalkanes' under the heading 'Nucleophilic substitutions'.

H_3C—CH_2—CH_2—O—$\overset{\overset{\displaystyle O}{\|}}{C}$—$CH_3$

Propylethanoate

↓ H_2O

H_3C—CH_2—CH_2—OH   +   HO—$\overset{\overset{\displaystyle O}{\|}}{C}$—$CH_3$

Propan-1-ol   Ethanoic acid

Ester hydrolysis (catalyst concentrated sulfuric acid)

H_3C—CH_2—CH_2—$\overset{\overset{\displaystyle O}{\|}}{C}$—$\overset{\overset{\displaystyle H}{|}}{N}$—H

Butanamide

↘ H_2O

H_3C—CH_2—CH_2—$\overset{\overset{\displaystyle O}{\|}}{C}$—OH   + NH_3

Butanoic acid

Hydrolysis of an amide

Properties of carboxylic acids

Aqueous solutions of carboxylic acids have a pH less than 7. Since they are weak acids, the pH is dependent on the concentration of the acid and the pKa value (see the 'Strong/weak acids and bases' section in 'Chemical equilibrium').

Carboxylic acids are capable of forming hydrogen bonds. This means that they have higher boiling points than alkanes of a similar molecular mass. In a concentrated carboxylic acid, dimers can form between two molecules due to the hydrogen bonding.

Dimers forming in a concentrated carboxylic acid

Small carboxylic acids are soluble in water since they can form hydrogen bonds with water molecules.

Reactions of carboxylic acids

Carboxylic acids behave as typical acids in aqueous solution and form salts by reacting with metals and bases including alkalis (soluble bases). They also undergo condensation reactions with alcohols to form esters (covered in the Alcohols section of this topic). Carboxylic acids react with amines to form amides and they can be reduced by using lithium aluminium hydride (LiAlH$_4$) to directly form primary alcohols due to the LiAlH$_4$ being such a powerful reducing agent. Sodium borohydride (NaBH$_4$) can also be used as a reducing agent.

Metal and carboxylic acid

An example of this reaction is that of magnesium and ethanoic acid to form the salt magnesium ethanoate and hydrogen gas.

$$Mg(s) + 2CH_3COO^-H^+(aq) \rightarrow H_2(g) + Mg^{2+}(CH_3COO^-)_2(aq)$$

Base and carboxylic acid

Sodium carbonate reacts with ethanoic acid to form the salt sodium ethanoate, water and carbon dioxide gas.

$$Na_2CO_3(s) + 2CH_3COOH(aq) \rightarrow CO_2(g) + H_2O(l) + 2Na^+CH_3COO^-(aq)$$

Condensation reactions to form esters

Please see the Alcohols section of this topic and Higher Chemistry.

Condensation reactions to form amides

Also covered at Higher Chemistry.

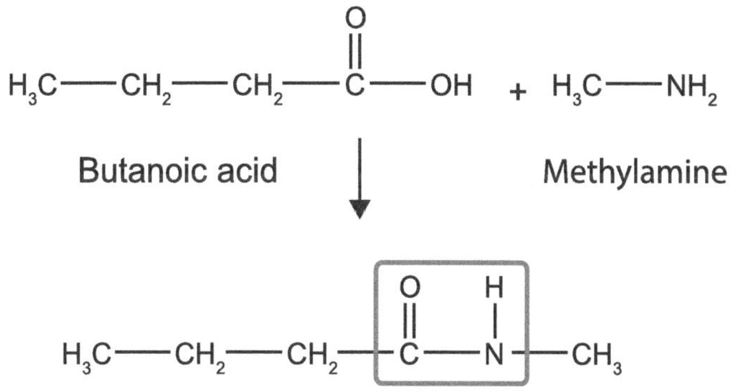

TOPIC 2. SYNTHESIS

$$H_3C-CH_2-CH_2-\overset{\overset{O}{\|}}{C}-OH \quad + \quad H_3C-NH_2$$

Butanoic acid　　　　　　　　　Methylamine

$$H_3C-CH_2-CH_2-\boxed{\overset{\overset{O}{\|}}{C}-\overset{\overset{H}{|}}{N}}-CH_3$$

An amide with the amide linkage highlighted, this can also be called a peptide link

Reaction with LiAlH₄ (or NaBH₄) to form primary alcohols

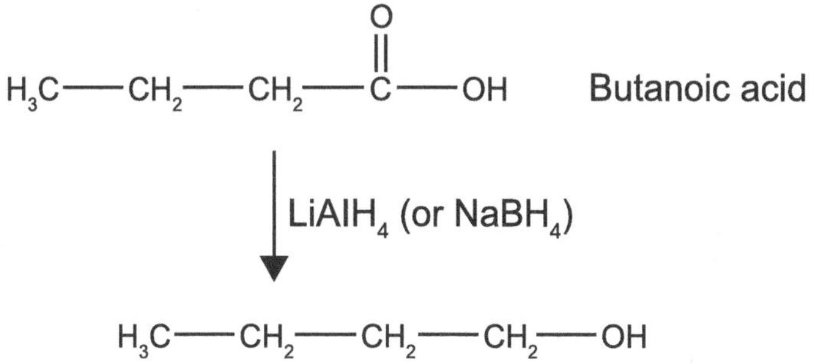

$$H_3C-CH_2-CH_2-\overset{\overset{O}{\|}}{C}-OH \quad \text{Butanoic acid}$$

$\downarrow$ LiAlH$_4$ (or NaBH$_4$)

$$H_3C-CH_2-CH_2-CH_2-OH$$

Butan-1-ol (primary alcohol)

© HERIOT-WATT UNIVERSITY

2.11 Amines

Amines are derived from ammonia where one or more of the hydrogen atoms have been replaced with an alkyl group. Amines are classified according to the number of alkyl groups attached to the nitrogen atom.

Primary amine where one of the hydrogens of ammonia has been replaced with an alkyl group.

$$H_3C-N\begin{matrix}H\\ \\H\end{matrix}$$

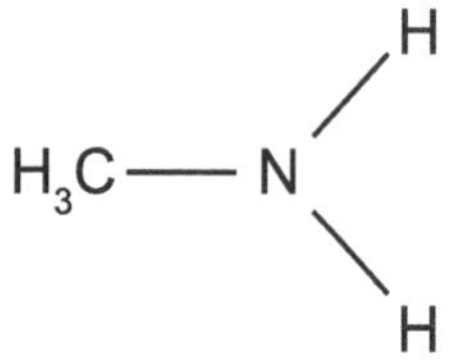

Primary amine

Secondary amine where two of the hydrogen atoms of ammonia have been replaced with an alkyl group. These do not necessarily have to be the same alkyl group.

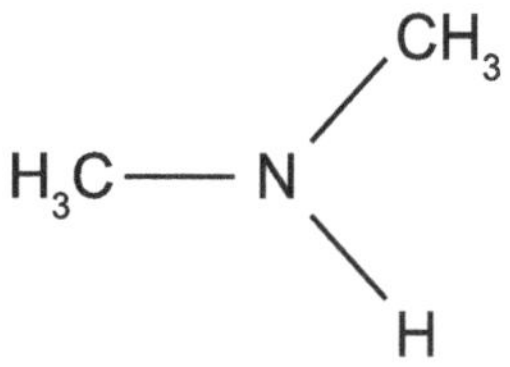

$$H_3C-N\begin{matrix}CH_3\\ \\H\end{matrix}$$

Secondary amine

Tertiary amine where all three hydrogen atoms of ammonia have been replaced with alkyl groups. Again these alkyl groups are not necessarily all the same.

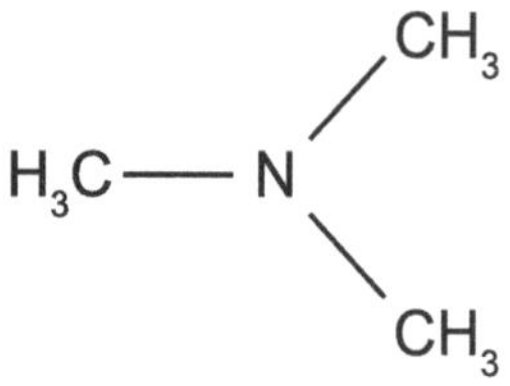

$$H_3C-N\begin{matrix}CH_3\\ \\CH_3\end{matrix}$$

Tertiary amine

TOPIC 2. SYNTHESIS

Naming amines

Amines are named by prefixing the word amine with the names of the alkyl groups attached to the nitrogen atom arranged in alphabetical order.

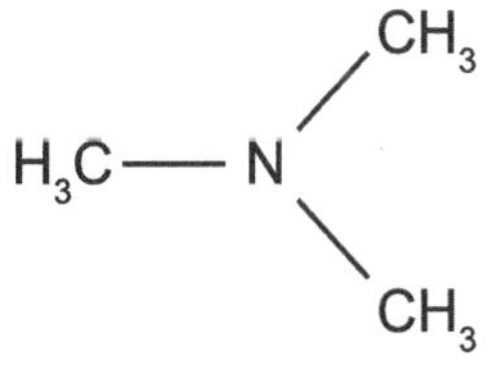

$$H_3C-N(CH_3)(CH_3)$$

Trimethylamine

$$H_3C-N(CH_2CH_3)(CH_3)$$

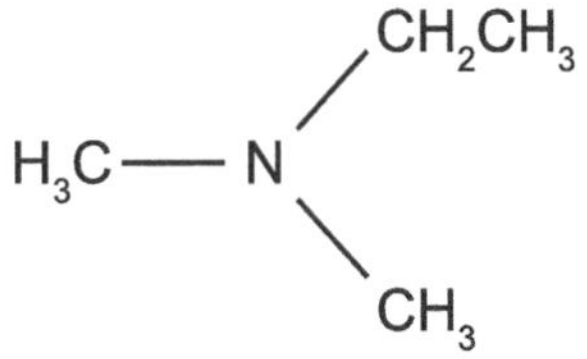

Ethyldimethylamine

Properties of amines

A polar N-H bond is found in primary and secondary amines therefore they have hydrogen bonding between their molecules. These do not occur in tertiary amines due to the lack of a hydrogen atom bonded directly to the highly electronegative nitrogen atom. This causes primary and secondary amines to have higher boiling points compared to their isomeric tertiary amines.

All types of amine can form hydrogen bonds with water and hence they are soluble in water. This is shown in the following diagram.

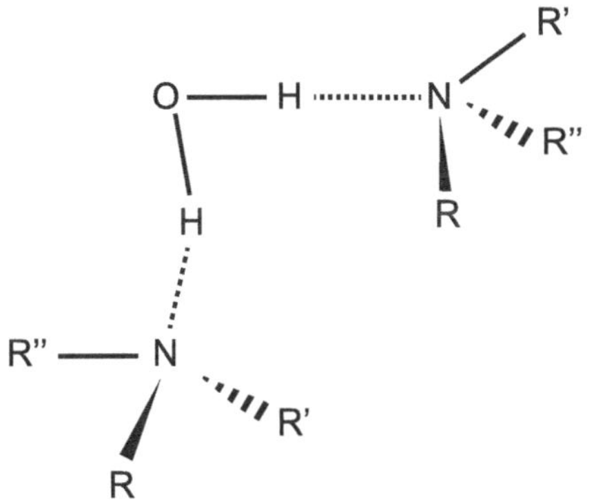

Amines are weak bases like ammonia and will dissociate slightly in aqueous solutions. The lone pair of electrons on the nitrogen in the amine molecule accepts a proton from the water molecule forming a alkylammonium ions and hydroxide ions. The hydroxide ions make the solution alkaline.

$$CH_3NH_2(aq) + H_2O(l) \rightleftharpoons CH_3NH_3^+(aq) + OH^-(aq)$$

Reactions of amines

Amines react with hydrochloric acid, sulfuric acid and nitric acid to form salts.

$$CH_3CH_2NH_2 + HNO_3 \rightarrow CH_3CH_2NH_3^+NO_3^- \text{ (ethylammonium nitrate)}$$

They also react with carboxylic acids to form salts. Amides are formed if these salts are heated losing water.

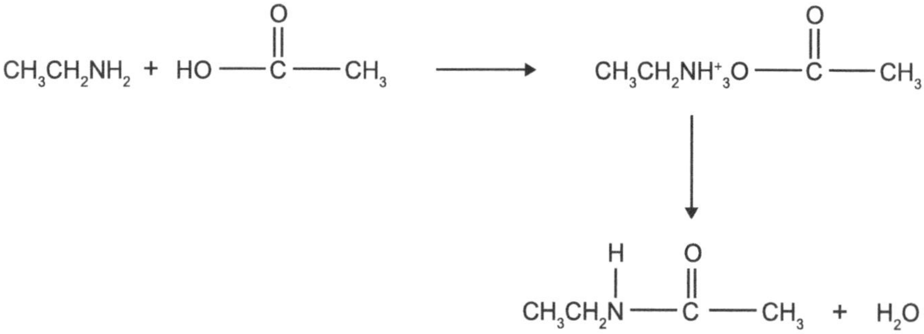

Reaction of an amine with a carboxylic acid to form a salt

2.12 Aromatic compounds

The simplest member of the class of compounds known as **aromatic** compounds is benzene. It has the molecular formula C_6H_6. In the planar benzene molecule each carbon atom is sp^2 hybridised and the three filled sp^2 hybrid orbitals form sigma bonds with a hydrogen atom and two neighbouring carbon atoms. This leaves an electron occupying a p orbital on each carbon atom. Each of these p orbitals overlaps side on with the two p orbitals on either side and a pi molecular orbital forms.

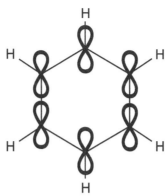

P orbitals above and below each carbon atom

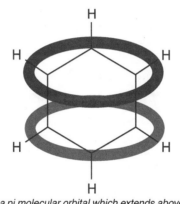

Overlap of p orbitals to form a pi molecular orbital which extends above and below the benzene ring

The six electrons that occupy the pi orbital are not tied to any one of the carbon atoms and are shared by all six. They are known as delocalised electrons. The structure is represented as follows:

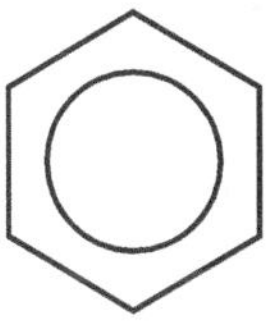

The following structure, called the Kekulé structure, wrongly illustrates that benzene has alternating single and double bonds. Historically, this was an accepted structure for benzene but it is not correct as benzene does not react with Br_2(aq) proving that it doesn't have C=C bonds. Analysis of the bonds shows that all six C-C bonds in the ring are equal length and energy, between those of C-C single bonds and C=C double bonds.

2.13 Reactions of benzene

The delocalised electrons give benzene an unusual stability and enable it to undergo substitution reactions rather than addition reactions which would disrupt the stability of the ring. Benzene is readily attacked by electrophiles, due to the high electron density of the delocalised ring system, through reactions such as alkylation, chlorination, nitration and sulfonation, which are all examples of **electrophilic substitution**.

When writing reaction mechanisms for aromatic systems, it is often useful to represent them as Kekulé structures (the localised arrangements). It should always be remembered however that this is not a true picture of the benzene molecule.

The electrophile attacks one of the carbons atoms in the ring, forming an intermediate ion with a positive charge that can be stabilised by delocalisation around the ring. The intermediate ion then loses H⁺ to regain aromatic character and restore the system.

As well as chlorination using aluminium chloride or iron (III) chloride, benzene can undergo bromination using iron (III) bromide.

The catalyst polarises the halogen molecule (here it is bromine) by accepting a pair of electrons from one atom and creating an electrophilic centre on the other atom.

$$\overset{\delta^+}{Br} \cdots\cdots \overset{\delta^-}{Br} \cdots\cdots FeBr_3$$

Electrophilic centre on bromine

This partially positive bromine atom attacks one of the carbons on the benzene and creates a carbocation which is stabilised by delocalisation on the ring. The intermediate then loses a

TOPIC 2. SYNTHESIS

hydrogen ion by heterolytic fission from the benzene, regaining the aromatic character and forming bromobenzene. The iron(III) bromide is regenerated. The same mechanism applies if chlorine is used with a suitable catalyst.

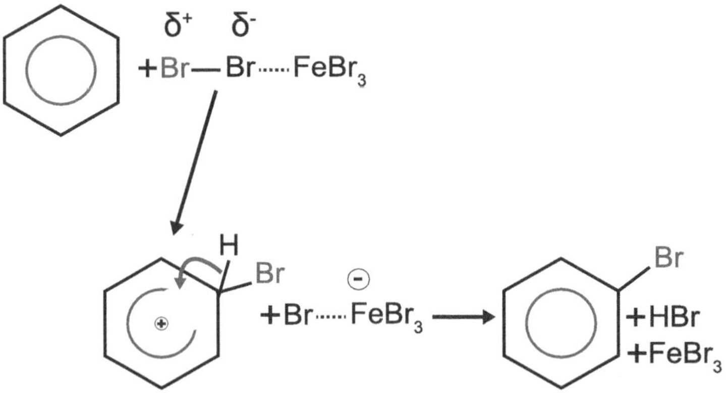

The reaction of chlorine with benzene

Answer these questions concerning the similar reaction of chlorine with benzene in the presence of aluminium(III) chloride.

Q1: What kind of reaction would take place?

a) Electrophilic addition
b) Nucleophilic substitution
c) Electrophilic substitution
d) Nucleophilic addition

....

Q2: What is generally the best name for the product?

a) Chlorophenyl
b) Chlorobenzene
c) Benzene chloride
d) Phenylchloride

....

Q3: What type of fission does the chlorine molecule undergo in this reaction?

a) Nuclear
b) Electrolytic
c) Homolytic
d) Heterolytic

....

Q4: What type of organic ion is created as an intermediate in this reaction?

Benzene will react with nitric acid when a mixture of concentrated nitric acid and concentrated sulfuric acid (known as a nitrating mixture) is used and the temperature is kept below 55°C. The nitrating mixture generates the nitronium ion, NO_2^+.

$$HNO_3 + 2H_2SO_4 \longrightarrow H_3O^+ + 2HSO_4^- + NO_2^+$$

Nitrobenzene

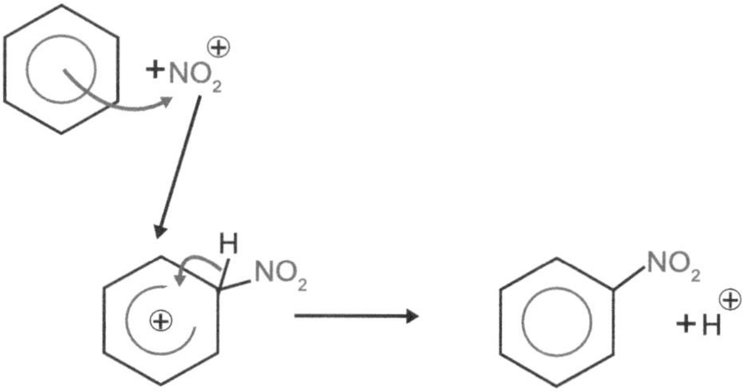

If the temperature is allowed to rise above 55°C, further substitution of the ring can take place and small amounts of the di- and tri- substituted compounds can result.

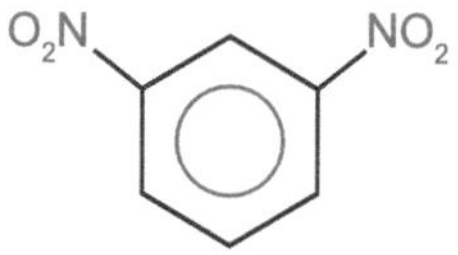

Structure x

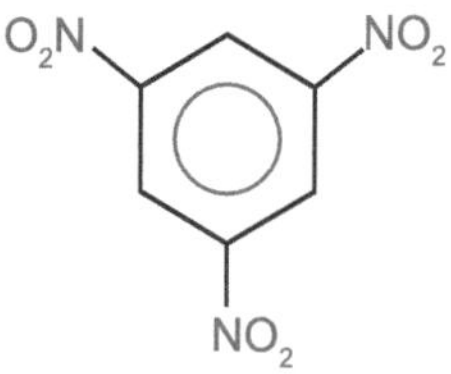

Di- and trinitrobenzene structures

Di- and trinitrobenzene structures

Q5: What type of reactant is the nitronium ion?

a) Carbanion
b) Nucleophile
c) Carbocation
d) Electrophile

TOPIC 2. SYNTHESIS

Q6: What type of intermediate is formed?

a) Free radical
b) Carbocation
c) Carbanion
d) Nitronium anion

....

Q7: Describe how the intermediate is stabilised. Try writing a sentence before checking your answer.

....

Q8: What name would you suggest for structure (x)?

a) 1,1-dinitrobenzene
b) 1,2-dinitrobenzene
c) 1,3-dinitrobenzene
d) 1,4-dinitrobenzene

Benzene will react with concentrated sulfuric acid if the reactants are heated together under reflux for several hours. If fuming sulfuric acid is used (sulfuric acid enriched with sulfur trioxide) under cold conditions $^+SO_3H$ substitutes onto the ring. This suggests that the active species is sulfur trioxide. The sulfur atom in the sulfur trioxide molecule carries a partial positive charge and can attack the benzene ring. The mechanism is the same as that shown for nitration.

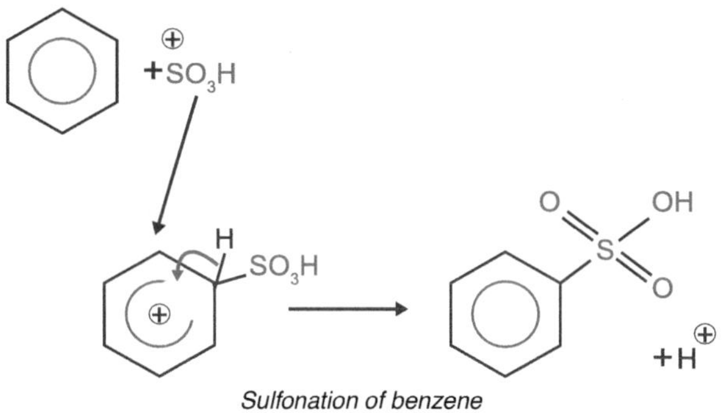

Sulfonation of benzene

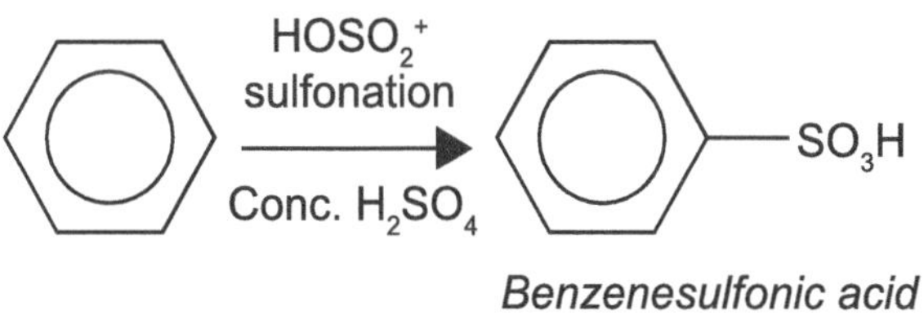

Benzenesulfonic acid

Benzene

Q9: In this reaction is the sulfur trioxide acting as a nucleophile or a electrophile?
....

Q10: Does it react with the benzene in an addition or a substitution reaction?

The aluminium chloride catalyst mentioned in the halogenation reactions can be used to increase the polarisation of halogen containing organic molecules like halogenoalkanes. This allows an electrophilic carbon atom to attack the benzene ring and builds up side-chains. The reaction is called a Friedel-Crafts reaction, after the scientists who discovered it.

$$\overset{\delta^+}{R} \cdots \overset{\delta^-}{Cl} \cdots AlCl_3$$

Electrophilic centre on alkyl group

The catalyst increases the polarity of the halogenoalkane producing the electrophilic centre that can attack the benzene ring. The carbocation formed is stabilised by delocalisation and the intermediate so formed regains its stability by loss of a hydrogen ion forming an alkylbenzene.

Where R- is an alkyl group such as methyl $-CH_3$

Methylbenzene

Methylbenzene

Q11: What word describes the role of the aluminium(III) chloride?

....

Q12: What type of fission does the halogenoalkane undergo in this reaction?

a) Nuclear
b) Homolytic
c) Heterolytic
d) Electrolytic

....

Q13: Name the chloroalkane necessary to produce ethylbenzene in a Friedel-Crafts reaction.

One or more H atoms can be substituted on the benzene ring which leads to a wide range of consumer products including many pharmaceutical drugs. (The benzene ring is usually drawn in a different manner in these structures).

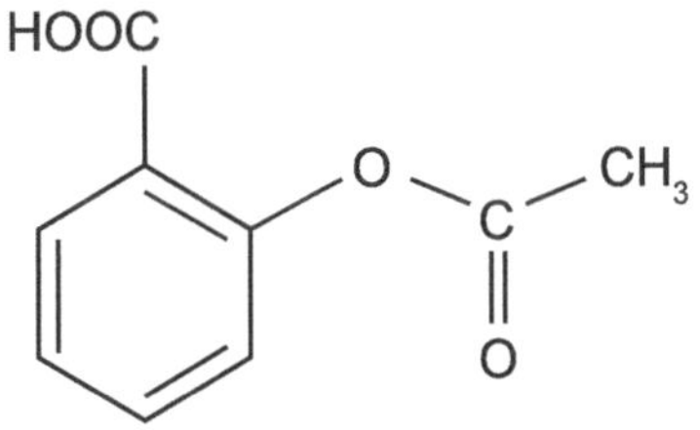

Structure of aspirin

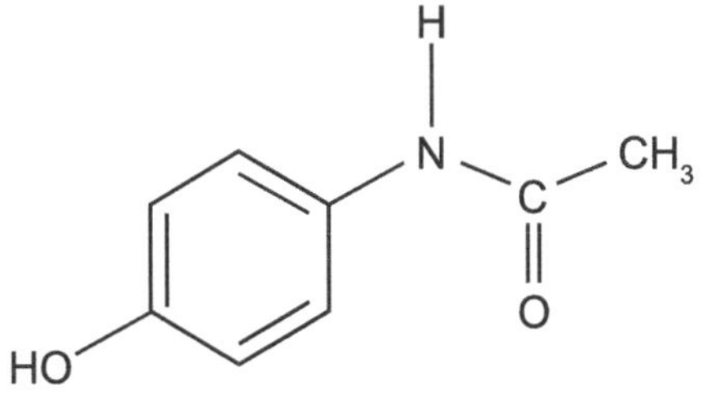

Structure of paracetamol

Phenyl

Where one of the hydrogen atoms has been substituted in a benzene ring it is known as the phenyl group (C_6H_5).

This molecule is known as phenol

2.14 Extra practice

Extra practice questions

Some possible reactions of propenal are shown as follows.

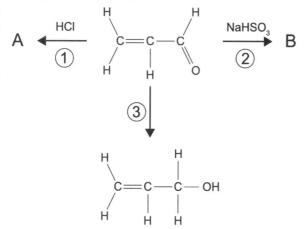

Q14: Draw a structural formula for compound **A** which is the major product formed during the reaction.

....

Q15: Which reagent could be used to carry out reaction **3**?

TOPIC 2. SYNTHESIS

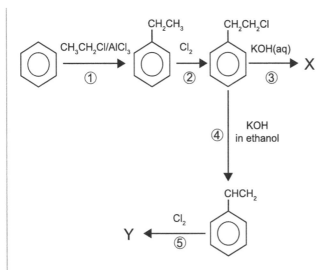

Q16: What type of reaction is taking place in step 1?

....

Q17: What experimental condition would be required in step 2?

....

Q18: Draw a structural formula for product **X**.

....

Q19: What type of reaction is taking place in step 4?

....

Q20: Draw a structural formula for product **Y**.

Consider the structure of lactic acid:

$$\text{H-}\underset{\underset{H}{|}}{\overset{\overset{H}{|}}{C}}\text{-}\underset{\underset{H}{|}}{\overset{\overset{OH}{|}}{C}}\text{-}\underset{OH}{\overset{O}{C}}$$
  1   2   3

Q21: What is the systematic name of lactic acid?

....

Q22: Lactic acid contains an aysmmetric carbon atom. Identify, and **explain**, which of the numbered carbon atoms is asymmetric.

....

© HERIOT-WATT UNIVERSITY

Q23: Lactic acid can be produced from ethanal by the reaction sequence as follows.

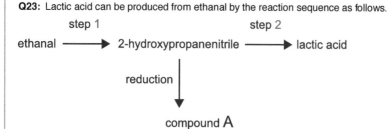

Which reagent could be used in **step 1**?

....

Q24: What type of reaction takes place in **step 2**?

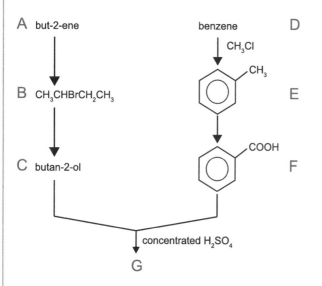

Q25: Explain why but-2-ene exhibits geometric isomerism yet its structural isomer but-1-ene does not.

....

Q26: But-2-ene undergoes electrophilic addition to form **B**. Draw a structure for the carbocation intermediate formed in this electrophilic addition reaction.

....

Q27: Name a reagent used to convert **B** to **C**.

....

TOPIC 2. SYNTHESIS

Q28: Name a catalyst required in converting **D** to **E**.

....

Q29: Draw a structural formula for ester **G**.

2.15 Summary

Summary

You should now be able to:

- interchange between and recognise the molecular, structural and skeletal formulae of various compounds.
- recognise and use different types of reaction in organic synthesis including substitution, addition, elimination, condensation, hydrolysis, oxidation, reduction.
- devise synthetic routes, with no more than three steps, from a given reactant to a final product.
- look at molecular structures and deduce the reactions it can undergo.

2.16 Resources

- LearnChemistry, Organic formulae (http://rsc.li/2az02xA)
- Chemguide, Free radical substitution (http://bit.ly/2azRAex)
- The 8 Types of Arrows In Organic Chemistry, Explained (http://bit.ly/2awwQWo)
- Chemistry in Context Laboratory Manual, fifth edition, Graham Hill and John Holman ISBN 0-17-448276-0
- Chemguide, Nucleophilic substitution (http://bit.ly/2ajukQb)
- Comparing the SN1 and SN2 Reactions (http://bit.ly/1Nwjiml)
- The Williamson Ether Synthesis (http://bit.ly/2aMcW7A)
- Chemguide, Hydrolysing Nitriles (http://bit.ly/2az2O5Q)

2.17 End of topic test

End of Topic 2 test

Q30:

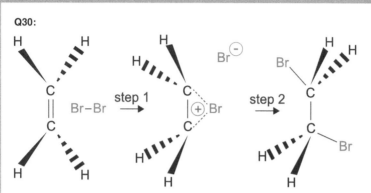

The two steps in the mechanism shown could be described as:

a) ethene acting as an nucleophile and Br⁻ acting as an nucleophile.
b) homolytic fission of the Br₂ followed by Br⁻ acting as a nucleophile.
c) ethene acting as an electrophile and Br⁻ acting as an electrophile.
d) ethene acting as an electrophile and Br⁻ acting as an nucleophile.

....

Q31: Name the missing compound in this reaction of propene.

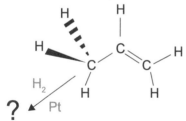

....

Q32: Name the missing compound in this reaction of propene.

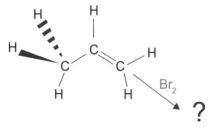

TOPIC 2. SYNTHESIS

Q33: What type of halogenoalkane is shown?

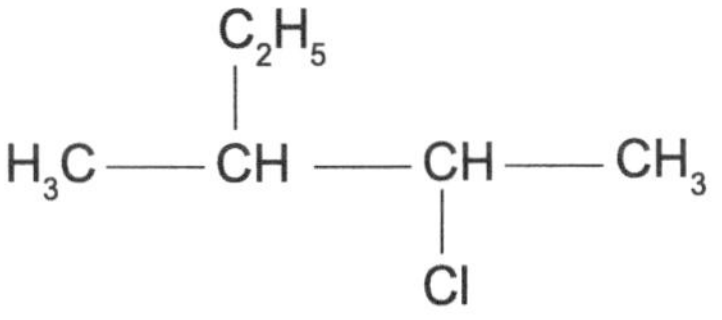

$$H_3C-CH(C_2H_5)-CH(Cl)-CH_3$$

a) Primary
b) Secondary
c) Tertiary
d) None of the above

Q34: Give the correct name for the compound shown.

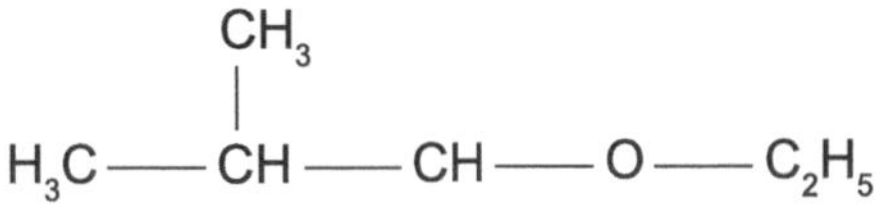

$$H_3C-CH(CH_3)-CH-O-C_2H_5$$

Q35: The diagram shows the structure of an alcohol. What type of alcohol is it?

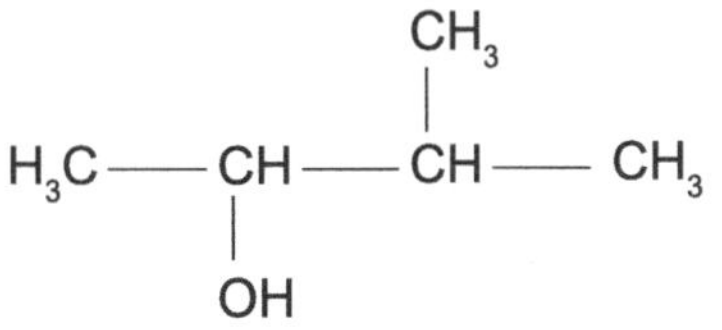

$$H_3C-CH(OH)-CH(CH_3)-CH_3$$

Q36: Which of the following could not be used to prepare this alcohol?

a) 2-methylbut-2-ene
b) 2-methylbut-1-ene
c) 3-methylbut-1-ene
d) 2-bromo-3-methylbutane

Esters can be prepared from alcohols by two alternative routes.

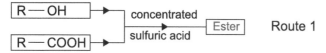 Route 1

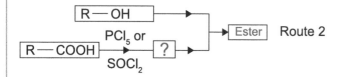 Route 2

Q37: What type of compound is missing in Route 2?

....

Q38: What is the disadvantage in using Route 1?

....

Q39: Which of the following shows the structure of the ester formed when propan-2-ol and ethanoic acid react?

1 $H_3C - CH_2 - \underset{\underset{O}{\|}}{C} - O - CH_2 - CH_3$

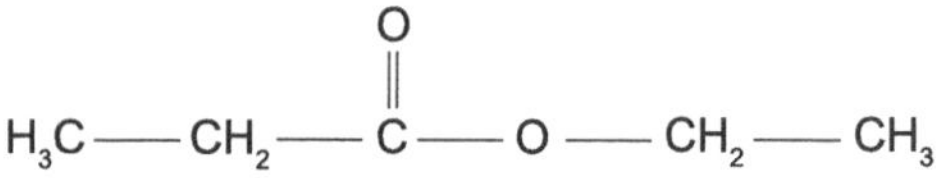

2 $H_3C - \underset{\underset{CH_3}{|}}{CH} - O - \underset{\underset{O}{\|}}{C} - CH_2 - CH_3$

3 $H_3C - \underset{\underset{CH_3}{|}}{CH} - O - \underset{\underset{O}{\|}}{C} - CH_3$

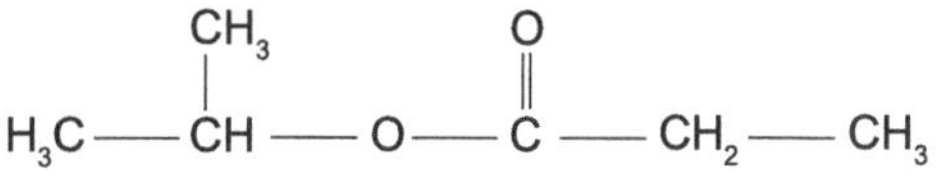

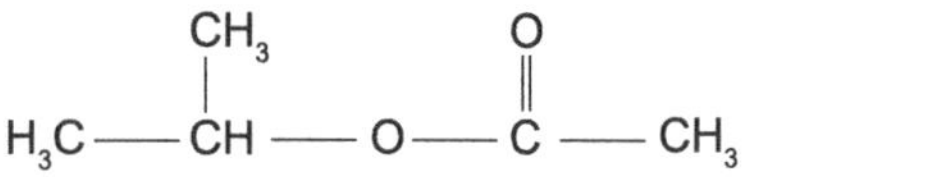

4 $H_3C - CH_2 - CH_2 - O - \underset{\underset{O}{\|}}{C} - CH_3$

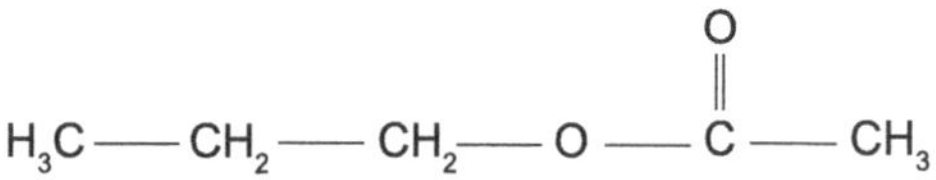

a) 1
b) 2
c) 3
d) 4

....

TOPIC 2. SYNTHESIS

Q40: Carboxylic acids are weak acids. Which of the following statements is true?

a) Hexanoic acid will be more soluble than propanoic acid.
b) Ethanol will have a lower value for pK_a than ethanoic acid.
c) The ethanoate ion is more stable than the ethoxide ion due to electron delocalisation.
d) In dilute aqueous ethanoic acid, hydrogen bonding between acid molecules produces dimers.

Propanoic acid can react with a number of different reagents as show in the following scheme.

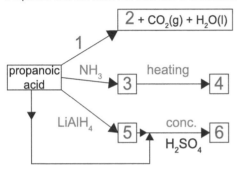

Q41: What could reagent **1** be?

a) $MgCO_3$
b) NaOH
c) CuO
d) H_2SO_4

....

Q42: What type of compound are **2** and **3**?

....

Q43: What type of compound is **4**?

....

Q44: Name compound **5**.

....

Q45: Name compound 6

....

Q46: Which of these statements describes the reactivity of the benzene ring? It:

a) resists electrophilic attack.
b) rapidly decolourises bromine water.
c) reacts rapidly with nucleophiles.
d) resists addition reactions.

© HERIOT-WATT UNIVERSITY

Q47: Which of these is a major product of the reaction between 2-chloropropane and benzene in the presence of aluminium(III) chloride?

a) 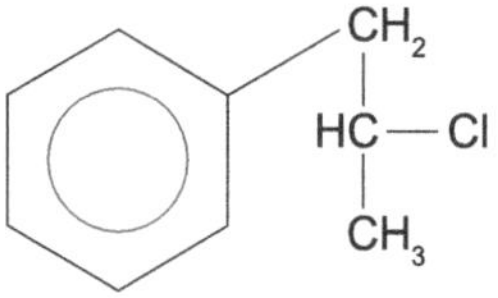 C6H5–CH2–CH(Cl)–CH3

b) 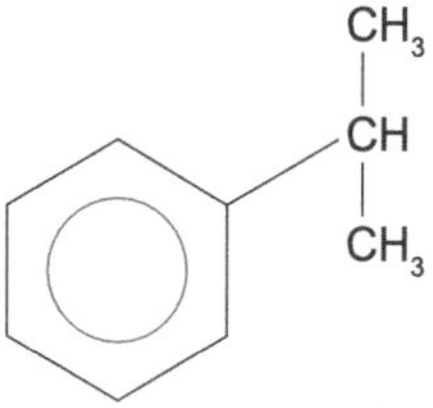 C6H5–CH(CH3)–CH3

c) 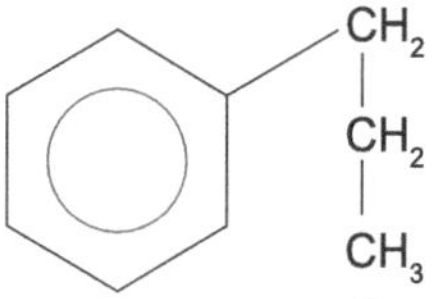 C6H5–CH2–CH2–CH3

d) 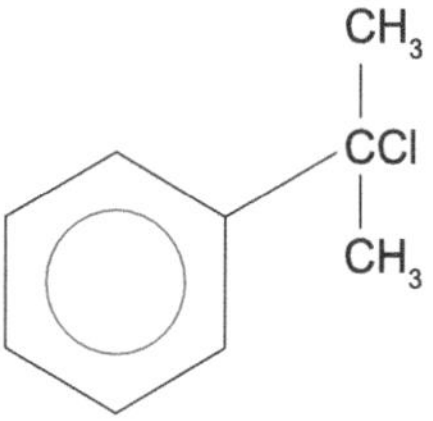 C6H5–C(Cl)(CH3)–CH3

Derivatives of benzene with more than one substituted side group are named by numbering the ring and defining the substituted group position. Example:

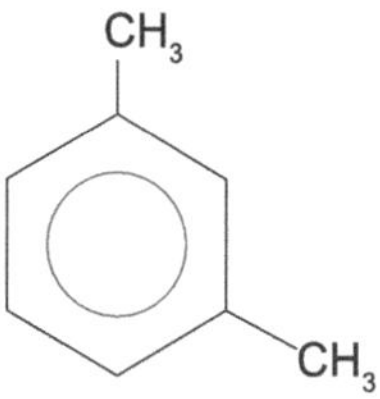

1,3-dimethylbenzene

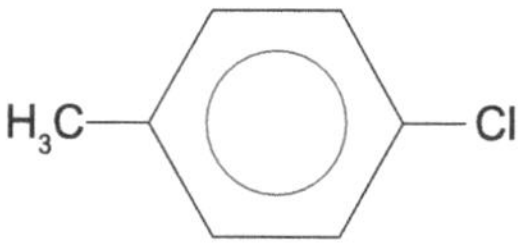

1-chloro-4-methylbenzene

Q48: Name the following molecule:

....

Q49: Name the following molecule:

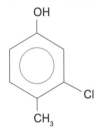

....

Q50: Suggest a name for a possible aromatic isomer of 1,3-dimethylbenzene

....

Q51: Which of the following reaction types are likely to be involved in the conversion of benzene to a typical detergent?

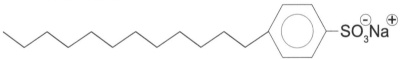

1. Addition
2. Alkylation
3. Condensation
4. Nitration
5. Polymerisation
6. Sulfonation

....

Q52: Which of these reactions would occur between benzene and a mixture of concentrated nitric acid and concentrated sulfuric acid?

a) Addition
b) Alkylation
c) Condensation
d) Nitration
e) Polymerisation
f) Sulfonation

Look at this molecule:

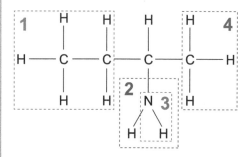

Q53: Which area is the functional group in this molecule?

a) 1
b) 2
c) 3
d) 4

Q54: To which class of amine does it belong?

a) Primary
b) Secondary
c) Tertiary
d) Aromatic

Q55: What name would you give the molecule?

a) Butan-2-amine
b) 2-methylpropanamine
c) 2-aminobutane
d) Ethylmethylamine

TOPIC 2. SYNTHESIS

Ethylamine reacts with nitric acid with both in the gaseous state and produces tiny white crystals.

Q56: What name would you give to the white crystals?

....

Q57: What type of reaction has taken place?

....

Q58: Which element is the electrophilic centre in this reaction?

....

Q59: Which element is the nucleophilic centre in this reaction?

....

Q60: Name a primary amine which should have a higher solubility in water than ethylamine.

Unit 4 Topic 3

Stereochemistry

Contents

- 3.1 Isomers 306
- 3.2 Geometric isomers 307
- 3.3 Optical isomers 310
- 3.4 Biological systems 312
- 3.5 Optical isomerism in medicines and other substances 314
- 3.6 Summary 315
- 3.7 Resources 315
- 3.8 End of topic test 316

Prerequisites

Before you begin this topic, you should know:

- that isomers are compounds that have the same molecular formula but differ in structural formulae. (National 5 and Higher Chemistry)

Learning objective

By the end of this topic, you should know:

- stereoisomers are isomers that have the same molecular formula but differ in structural formulae (a different spatial arrangement of their atoms);
- geometric isomers are stereoiosmers where there is a lack of rotation around one of the bonds mostly a C=C;
- these isomers are labelled cis and trans dependent on whether the substitutes are on the same or different sides of the C=C;
- optical isomers are non-superimposable mirror images of asymmetric molecules and are referred to as chiral molecules or enantiomers;
- isomers can often have very different physical or chemical properties from each other.

3.1 Isomers

Molecules which have the same molecular formula but differ in the way their atoms are arranged (structural formula) are called isomers. They often have very different physical and chemical properties from each other. There are two ways that atoms can be arranged in each isomer, one being that the atoms are bonded together in a different order. These are called structural isomers.

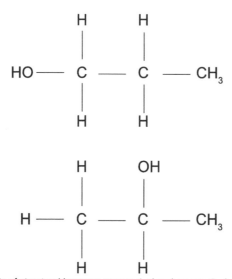

An example of structural isomers: propan-1-ol and propan-2-ol respectively

The second way involves the atoms being bonded in the same order but the arrangement of the atoms in space is different for each isomer. These are called stereoisomers and there are two different types we will study: geometrical and optical isomers.

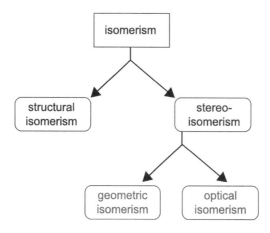

3.2 Geometric isomers

Geometric isomers can be illustrated by looking at the alkene but-2-ene.

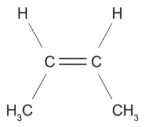

It contains a carbon to carbon double bond and the arrangement of atoms and bonds is planar with all the bond angles at 120°. The bonds are fixed in relation to each other meaning it is impossible to rotate one end of the alkene molecule around the carbon carbon double bond while the other end is fixed. This is why alkenes can exhibit geometric isomerism.

It would be advantageous at this point to make a model of but-2-ene using molymods to show the lack of rotation around the carbon to carbon bond.

There are two possible geometric isomers of but-2-ene and they are referred to as the **cis** isomer and the **trans** isomer.

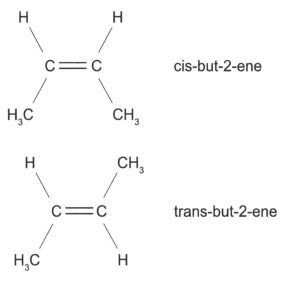

For geometric isomers cis means 'on the same side' and you can see from the diagram that the two methyl groups are on the same side of the carbon to carbon double bond. Trans means 'on different sides' and you can again see from the diagram that the methyl groups are on different sides of the carbon to carbon double bond. These two isomers have distinct physical properties; the melting point of the cis isomer is -139°C while the melting point of the trans isomer is -105°C.

There are also geometric isomers found in but-2-enedioic acid.

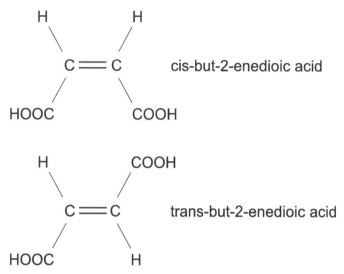

These isomers have differing physical properties but also have a chemical difference too in that the cis form is readily dehydrated. This is not possible in the trans isomer due to the carboxyl groups being on opposite sides of the carbon to carbon bond which is not a suitable orientation to undergo such a reaction. The melting point of the cis isomer is 135°C whereas the trans isomer has a melting point of 287°C again showing the effect on physical properties of isomers.

Geometric isomers can only occur in organic molecules which have two different groups attached to each of the carbon atoms of the double bond, in addition to the carbon to carbon double bond. For example propene would not exhibit geometric isomerism.

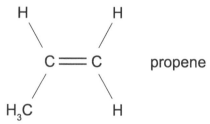

propene has two identical H atoms attached to one of the carbon atoms from the carbon to carbon bond

Geometric isomerism is most commonly found in organic molecules containing a C=C bond but it can also be found in saturated rings where rotation about the C-C single bonds is restricted.

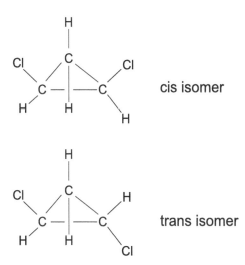

1,2-dichlorocyclopropane

Non-organic compounds can also exhibit geometric isomerism. For example the anti-cancer drug cisplatin.

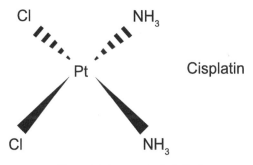

The trans-isomer has no comparable useful pharmacological effect.

3.3 Optical isomers

In the same way as your left hand and your right-hand are mirror images of each other, many chemical compounds can exist in two mirror image forms. Right and left hands cannot be superimposed on top of each other so that all the fingers coincide and are therefore not identical. A right hand glove does not fit a left hand and is said to be **chiral**.

The hands and the compounds have no centre of symmetry, plane of symmetry or axis of symmetry. Chirality arises from a lack of symmetry. Lack of symmetry is called asymmetry.

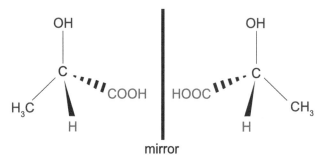

tetrahedral arrangement of an isomer showing the four different groups attached

two optical isomers of lactic acid: the mirror images are non-superimposable

Optical isomers are identical in all physical properties except their effect on plane polarised light, i.e. they exhibit optical activity. When plane polarised light is passed through a solution containing one optical isomer, the plane is rotated through a certain angle. If you have a solution of the other optical isomer at the same concentration the plane of polarised light is rotated by exactly the same angle but in the opposite direction. So if one isomer rotated the plane of polarised light by +60° (clockwise direction) the other isomer would rotate it by -60° (anticlockwise direction).

If you have a mixture of equimolar optical isomers it would have no effect on plane-polarised light since the rotational effect of one isomer would be cancelled out by the opposite rotational effect of the other. This mixture is optically inactive and is known as a **racemic mixture**.

Optical isomers of substances can be labelled R or S. The symbol R comes from the Latin rectus for right, and S from the Latin sinister for left. The four different groups are numbered in order of priority and if the order 1,2,3,4 is in a clockwise direction we call the **enantiomer** R. If they are in that order in an anticlockwise direction then we call the enantiomer S. The order of priority is determined by the element directly bonded to the chiral centre, the one with the highest atomic mass being numbered 1. If two groups have the same element directly bonded to the chiral centre then the next element bonded to that is considered and so on.

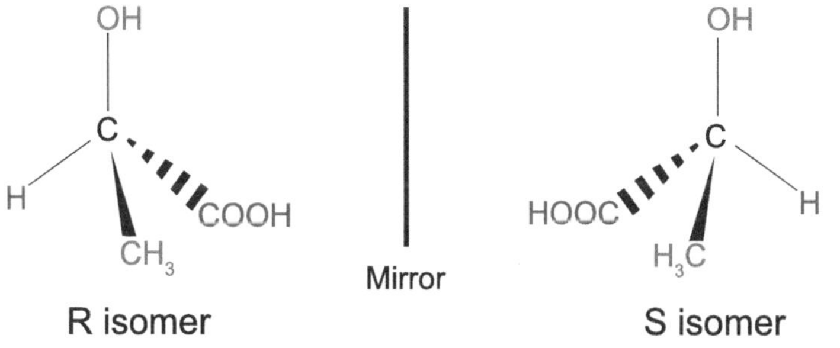

TOPIC 3. STEREOCHEMISTRY

```
         OH                                OH
         |                                 |
         C ⋯                             ⋯ C
       ⁄   ⁇  ⁇                       ⁇  ⁇   ⁍
      H     COOH               HOOC           H
           CH₃                        H₃C
        R isomer      Mirror        S isomer
```

Priority of groups

1 = OH
2 = COOH
3 = CH₃
4 = H

Optical isomerism is immensely important in biological systems. In most biological systems only one optical isomer of each organic compound is usually present. For example, when the amino acid alanine is synthesised in the laboratory, a mixture of the two possible isomers (a racemic mixture) is produced. When alanine is isolated from living cells, only one of the two forms is seen.

The proteins in our bodies are built up using only one of the enantiomeric forms of amino acids. Only the isomer of alanine with structure 2 (Figure 7.19) occurs naturally in organisms such as humans. Since amino acids are the monomers in proteins such as enzymes, enzymes themselves will be chiral. The active site of an enzyme will only be able to operate on one type of optical isomer. An interesting example of this is the action of penicillin, which functions by preventing the formation of peptide links involving the enantiomer of alanine present in the cell walls of bacteria. This enantiomer is not found in humans. Penicillin can therefore attack and kill bacteria but not harm the human host, as this enantiomer is not present in human cells.

3.4 Biological systems

Optical isomerism is immensely important in biological systems. In most biological systems only one optical isomer of each organic compound is usually present. For example, when the amino acid alanine is synthesised in the laboratory, a mixture of the two possible isomers (a racemic mixture) is produced. When alanine is isolated from living cells, only one of the two forms is seen.

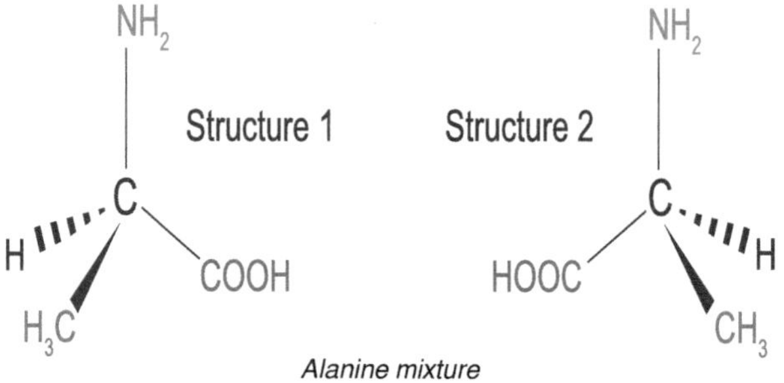

Alanine mixture

Optical isomers of alanine Go online

This illustration shows one of the enantiomers of alanine, drawn in a different way.

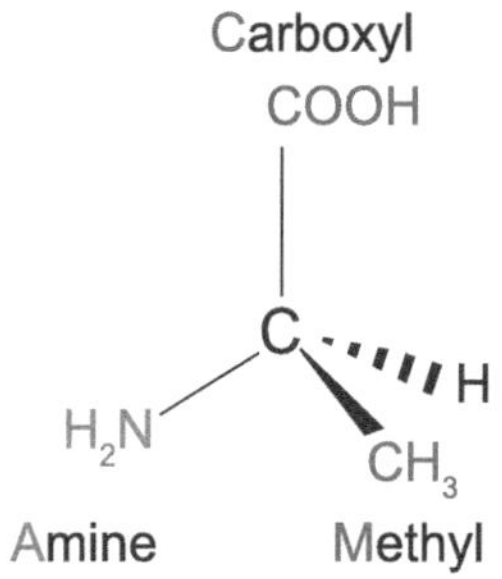

The carbon atom surrounded by four different groups is the **chiral centre**. As well as the hydrogen atom it has an amine group, a carboxyl group and a methyl group.

Q1: Compare the previous diagram with the following structures. Which structure is it?

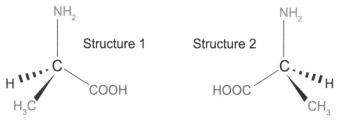

a) Structure 1
b) Structure 2

TOPIC 3. STEREOCHEMISTRY

If you have access to molecular models, you should build the molecule on the left of Picture 1 by starting with a carbon and using different coloured spheres for the four groups. (The suggested colours would be red for carboxyl, blue for amine, yellow for methyl and white for hydrogen.) Try to build the other enantiomer as the exercise progresses.

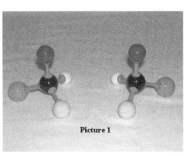

Picture 1

Picture 2

Picture 1 shows two isomers side by side and when one of them is taken away and replaced with a mirror, the two images in Picture 2 look the same as in Picture 1.

Picture 3

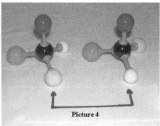

Picture 4

In Picture 3, the isomer that was removed is placed behind the mirror, showing that it is identical to the image. Picture 4 points out how the two isomers are non-superimposable because even with the two substituent groups on the left of each isomer lined up, the two arrowed groups do not match.

Q2: The molecules shown in Picture 4 are:

a) superimposable.
b) non-superimposable.

....

Q3: The two molecules are said to be:

a) symmetric.
b) asymmetric.

....

Q4: The two molecules can be said to be:

a) identical.
b) chiral.

3.5 Optical isomerism in medicines and other substances

Many medicines are produced as a mixture of enantiomers, only one of which is pharmacologically active, as it can prove costly to separate the isomers. One of the enantiomers of salbutamol used in the treatment of asthma is 68 times more effective than the other. Great care has to be taken when using drugs with enantiomeric forms as this has led to tragedy in the past, specifically in the case of thalidomide. A mixture of the isomers was used to treat nausea during pregnancy, and one enantiomer, which was thought to be inactive, turned out to cause damage to the unborn child. Many handicapped babies were born before the drug was recognised as being responsible. Screening of pharmaceuticals has to be very thorough. Regulations were tightened significantly after the thalidomide tragedy to ensure that both enantiomeric forms of chiral drugs are tested.

Structure of thalidomide

Another example where optical isomers are different is in limonene. One isomer of limonene smells of oranges, the other of lemons. Note the placement of the chiral carbon in the structure of limonene diagrams as follows.

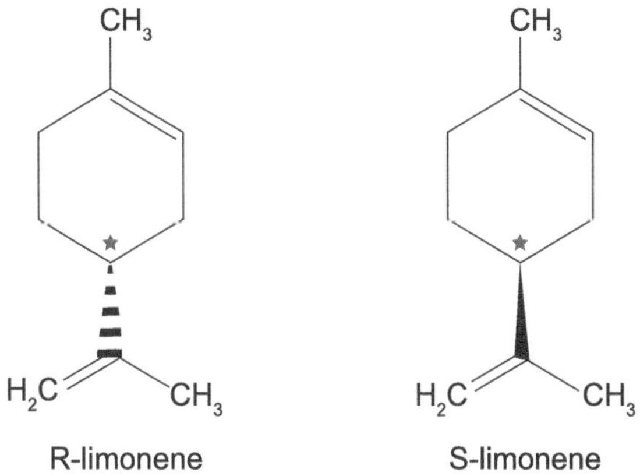

R-limonene   S-limonene

Structure of limonene

3.6 Summary

Summary

You should now be able to state that:

- stereoisomers are isomers that have the same molecular formula but differ in structural formulae.
- geometric isomers are stereoiosmers where there is a lack of rotation around one of the bonds mostly a C=C.
- these isomers are labelled cis and trans dependent on whether the substitutes are on the same or different sides of the C=C.
- optical isomers are non-superimposable mirror images of asymmetric molecules and are referred to as chiral molecules or enantiomers.
- isomers can often have very different physical or chemical properties from each other.

3.7 Resources

- LearnChemistry, Introduction to Chirality (http://rsc.li/2aiQ3ub)

3.8 End of topic test

End of Topic 3 test Go online

Q5: Which of the following could **not** exist in isomeric forms?

a) $C_2H_4Cl_2$
b) C_3H_6
c) C_3H_7Br
d) C_2F_4

....

Q6: Which of the following pairs represent the same chemical substance?

a) $H_2C{=}CH{-}CH_2{-}CH_3$ and $H_3C{-}CH{=}CH{-}CH_3$

b) and

c) and

d) and

....

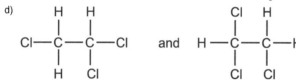

Q7: Which of these alcohols exists as optical isomers?

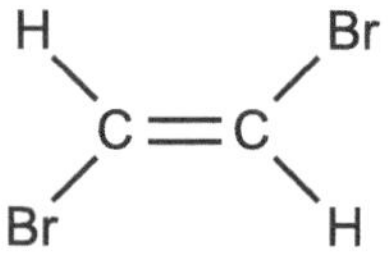

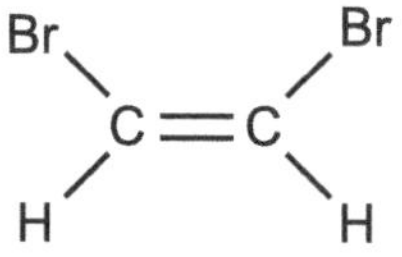

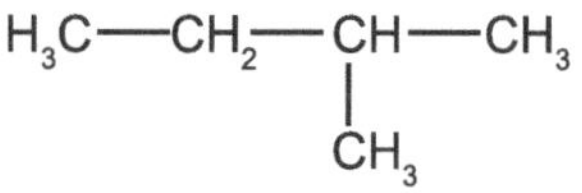

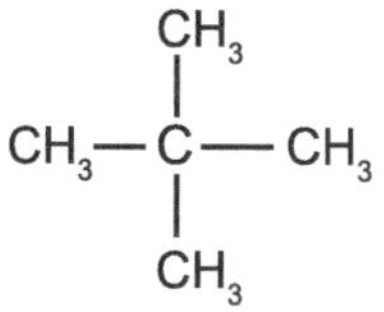

a) Propan-1-ol
b) Propan-2-ol
c) Butan-2-ol
d) Butan-1-ol

....

Q8: Which of these will have enantiomeric forms?

a) 1,2-dibromoethane
b) 1-chloroethanol
c) 1,2-bromochloroethene
d) 1-chloroethene

Take a look at these brominated alkenes:

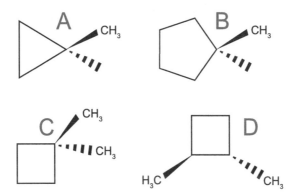

Q9: Which one contains an isomer of bromocyclopropane?

....

Q10: Which two brominated alkenes contain a molecule which is one of a cis/trans pair?

....

Q11: Which molecule will be the **least** polar?

Look at the structure of these cycloalkanes:

Q12: Which of the cycloalkanes shows cis/trans isomerism?

....

Q13: Which of the cycloalkanes is one of an enantiomeric pair?

....

Q14: Which of these amino acids does **not** have a chiral centre?

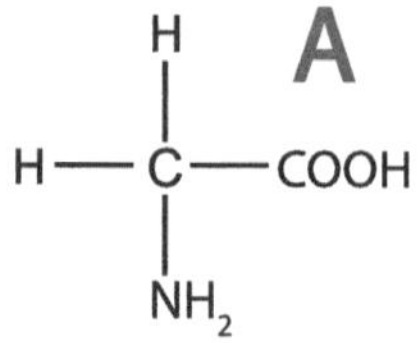

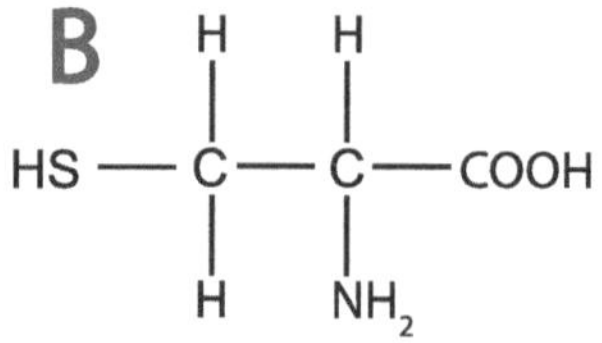

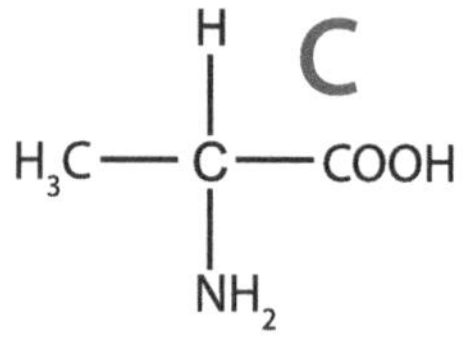

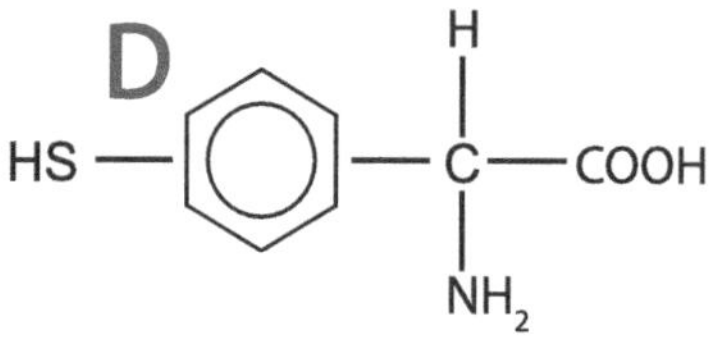

A: H—C(H)(NH₂)—COOH

B: HS—C(H)(H)—C(H)(NH₂)—COOH

C: H₃C—C(H)(NH₂)—COOH

D: HS—C₆H₄—C(H)(NH₂)—COOH

....

Q15: The part of a polarimeter capable of rotating a beam of polarised light is called the:
a) polariser.
b) analyser.
c) sample tube of chiral molecules.
d) light source.

....

Q16: Thalidomide is a drug which has optical isomers. Which atom is the chiral centre?

(Structure with labelled atoms W, X, Y, Z)

....

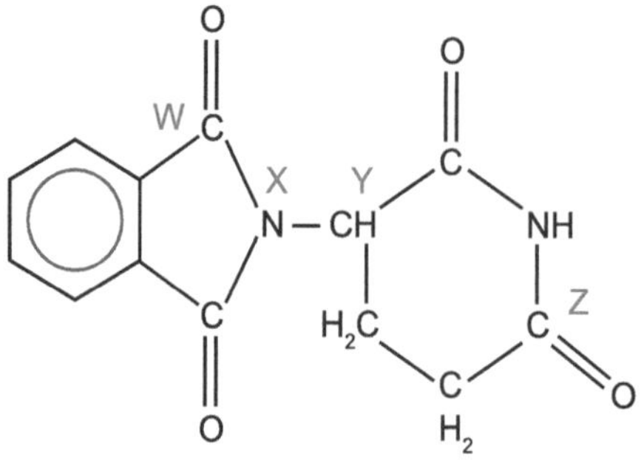

Q17: Carvone has a chiral centre. One form smells like spearmint and the other form smells like caraway seeds. Which atom is the chiral centre?

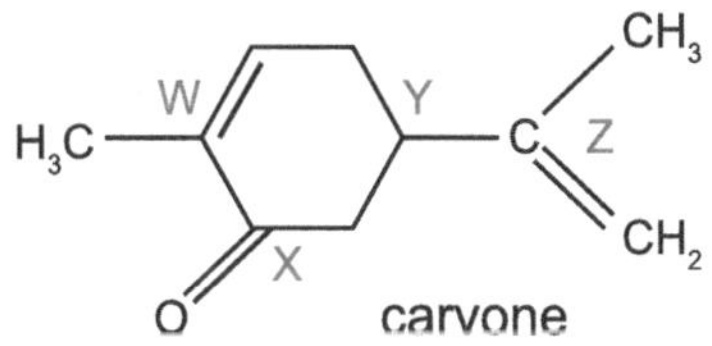

TOPIC 3. STEREOCHEMISTRY

carvone (structure with labels W, X, Y, Z, H₃C, CH₃, CH₂, O)

Q18: Carvone can undergo addition reactions with hydrogen molecules. How many moles of hydrogen would add onto one mole of carvone?

a) One
b) Two
c) Three
d) Four

Q19: What other word could be used to replace the phrase **optical isomers** of carvone?

This grid shows four compounds which all occur naturally in the fragrances of particular plants. A is geraniol (roses), B is nerol (bergamot), C is linalool (lavender) and D is citronellol (geraniums).

A — Geraniol (roses)
B — Nerol (bergamot)
C — Linalool (lavender)
D — Citronellol (geraniums)

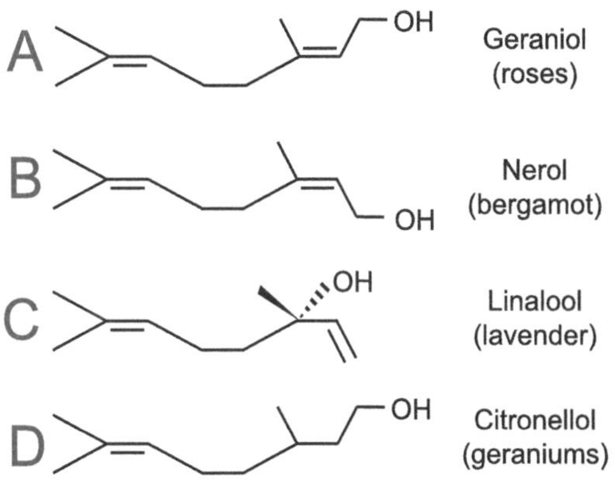

Q20: What is the relationship between A and B?

a) Structural isomers
b) Geometric isomers
c) Optical isomers
d) Not isomers

© HERIOT-WATT UNIVERSITY

Q21: What is the relationship between A and C?

a) Structural isomers
b) Geometric isomers
c) Optical isomers
d) Not isomers

....

Q22: Which compound (or compounds) could produce compound D on hydrogenation?

a) A only
b) B only
c) A and B
d) A, B and C

....

Q23: Which of the following contains a chiral entity?

a) C bonded to Cl, Cl, Br, I

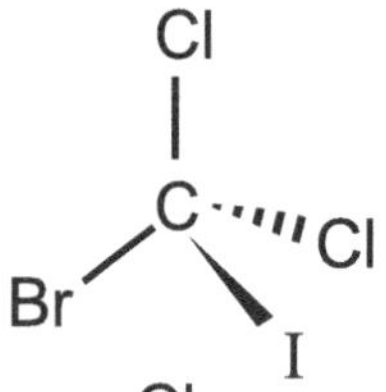

b) C bonded to Cl, Cl, H, CH_3

c) C bonded to Br, I, H, Cl

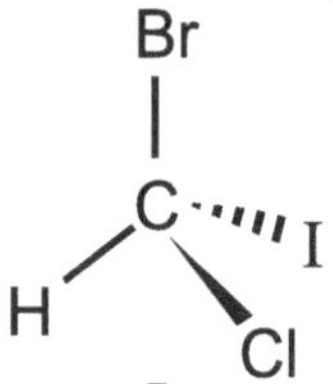

d) C bonded to I, Cl, Cl, Br

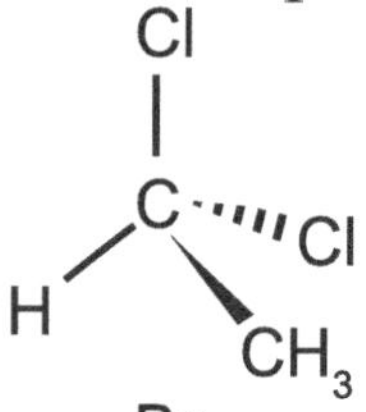

....

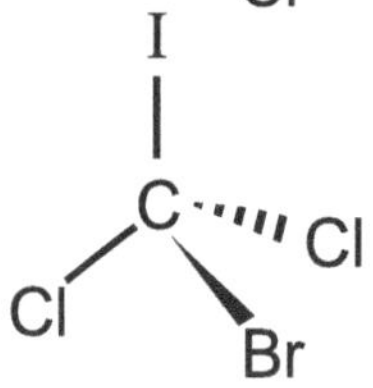

Q24: Which of the following contain identical entities?

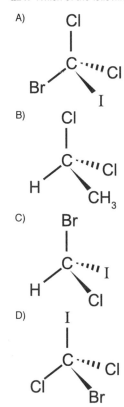

Unit 4 Topic 4

Experimental determination of structure

Contents

4.1 Elemental microanalysis 324
4.2 Mass spectrometry 325
4.3 Infrared spectroscopy 332
4.4 Proton nuclear magnetic resonance (NMR) spectroscopy 337
4.5 Absorption of visible light 347
4.6 Chromophores 348
4.7 Summary 350
4.8 Resources 350
4.9 End of topic test 351

Learning objective

By the end of this topic, you should be able to:

- explain that elemental microanalysis can be used to determine the masses of C, H, O, S and N in a sample of an organic compound in order to determine its empirical formula;
- work out the empirical formula from data given;
- explain that mass spectrometry can be used to determine the accurate molecular mass and structural features of an organic compound;
- explain infra-red spectroscopy can be used to identify certain functional groups in an organic compound and work out which compound is responsible for a spectra by identifying which functional groups are responsible for peaks;
- proton nuclear magnetic resonance spectroscopy (proton NMR) can give information about the different environments of hydrogen atoms in an organic molecule, and about how many hydrogen atoms there are in each of these environments;
- be able again to identify which compound is responsible for a spectra and be able to draw low resolution NMR spectra;
- explain how absorption of visible light by organic molecules occurs;
- state that the chromophore is the group of atoms within a molecule which is responsible for the absorption of light in the visible region of the spectrum.

4.1 Elemental microanalysis

Elemental microanalysis (combustion analysis) can be used to determine the empirical formula of an organic compound. The empirical formula shows the simplest whole number ratio of the different atoms that are present in a compound. If we look at propene for example then the molecular formula is C_3H_6 and hence its C:H ratio is 3:6. This simplifies to 1:2 and hence the empirical formula would be CH_2.

To carry out elemental microanalysis in modern combustion analysers a tiny sample (2 mg approximately) is accurately weighed and oxidised at a high temperature in an atmosphere of oxygen. This produces a mixture of gases SO_2, N_2, CO_2 and H_2O which are separated by gas chromatography and the mass of each component measured using a thermal conductivity detector. By converting the masses of each product gas into the masses of the original element (see the following example) and subtracting each of these from the original mass of the sample, the mass of oxygen in the compound can be obtained. The empirical formula is then determined from the calculated element masses using the following method.

Calculating the empirical formula

An antibiotic contains C, H, N, S and O. Combustion of 0.3442 g of the compound in excess oxygen yielded 0.528 g CO_2, 0.144 g H_2O, 0.128 g SO_2 and 0.056 g N_2. What is the empirical formula of the antibiotic?

1 mol CO_2 (44.0 g) contains 1 mole of C (12.0 g)
Mass of C in sample = 0.528 × 12 / 44 = 0.144 g

1 mol H_2O (18.0 g) contains 1 mole of H_2 (2 g)
Mass of H in sample = 0.144 × 2 / 18 = 0.016 g

1 mol SO_2 (64.1 g) contains 1 mole of S (32.1 g)
Mass of S in sample = 0.128 × 32.1 / 64.1 = 0.064 g

1 mol N_2 (28.0 g) contains 1 mole of N_2 (28.0 g)
Mass of N in sample = 0.056 × 28 / 28 = 0.056 g

Mass of O in sample = 0.3442 - (0.144 + 0.016 + 0.064 + 0.056) = 0.064 g

Element	C	H	S	N	O
Mass (g)	0.144	0.016	0.064	0.056	0.064
Number of moles	0.144 / 12 = 0.012	0.016 / 1 = 0.016	0.064 / 32.1 = 0.002	0.056 / 14 = 0.004	0.064 / 16 = 0.004
Mole ratio	0.012 / 0.002 = 6	0.016 / 0.002 = 8	0.002 / 0.002 = 1	0.004 / 0.002 = 2	0.004 / 0.002 = 2

Empirical formula is therefore $C_6H_8SN_2O_2$

You can also work out the empirical formula by knowing the percentage of each element in the compound.

A white solid is found to contain 66.67% carbon, 7.407% hydrogen and 25.936% nitrogen. What is the empirical formula of the solid?

Element	C	H	N
Percentage %	66.670	7.407	25.936
Number of moles	66.670 / 12 = 5.555	7.407 / 1 = 7.401	25.936 / 14 = 1.853
Mole ratio	5.555 / 1.853 = 3	7.407 / 1.853 = 4	1.853 / 1.853 = 1

Empirical formula is therefore C_3H_4N. The formula mass is reported to be 108 g and the molecular formula can be worked out from this.

Empirical formula = C_3H_4N = (3 × 12) + (4 × 1) + 14 = 54 g
Therefore the molecular formula is twice the empirical formula to make the mass 108 g = $C_6H_8N_2$.

Calculate the empirical formula

Q1: A liquid is found to contain 52.17% carbon, 13.04% hydrogen and the remainder is made up of oxygen. Work out the empirical formula of the liquid.

4.2 Mass spectrometry

Mass spectroscopy is a technique used in determining the accurate molecular mass of an organic compound. It can also determine structural features. A tiny sample (1×10^{-4} g approximately) of the unknown compound is vaporised and injected into the mass spectrometer. High-energy electrons bombard it with enough energy to knock electrons out of the molecules which are ionised and broken into smaller ion fragments.

The instrument is divided into four main sections:

1. *The inlet system.* This is maintained at a high temperature, so that any sample introduced will be vaporised rapidly. The interior of a mass spectrometer must be maintained at high vacuum to minimise collisions, so the injector and the rest of the instrument are connected to vacuum pumps.

2. *The ion source.* Some of the vaporised sample enters the ionisation chamber where it is bombarded with electrons. Provided the energy of the collision is greater than the molecular ionisation energy, some positive ions are produced from molecules in the sample material. Some of these molecular ions (sometimes called **parent ions**) will contain sufficient energy to break bonds, producing a range of fragments, some of which may also be positive ions (daughter ions).

$$M + e^- \rightarrow M^+ + 2e^-$$

The parent ions and ion fragments from the source are accelerated and focussed into a thin, fast moving ion beam by passing through a series of focusing slits.

3. *The analyser.* This stream of ions is then passed through a powerful magnetic field, where the ions experience forces that cause them to adopt curved trajectories.
Ions with the same charge (say, 1+) will experience the same force perpendicular to their motion, but the actual path of the ion will depend on its mass. Heavier ions will be deflected less than lighter ones.

4. *The detector.* Ions with the same mass/charge (m/z) ratio will have the same trajectory and can be counted by a detector. As the magnetic field strength is altered ions with different m/z ratios will enter the detector so that a graph of abundance (the ion count) against m/z values can be constructed. This is the mass spectrum of the sample compound(s).

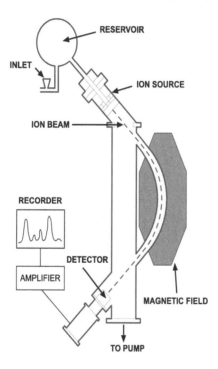

You can try a simple experiment at home. Get someone to roll golf balls and table tennis balls (heavy and light ions!) across a table. As the balls pass you, blow gently across their path. The two types of ball are about the same size, will experience about the same deflecting force from your breath, but will have different amounts of deflection owing to their different masses.

The following mass spectrum is for benzoic acid. The height of the vertical lines represents the relative abundance of ions of a particular m/z (mass/charge) ratio. The most abundant peak (the tallest one) is called the base peak and is given the abundance of 100%. The percentages of the other peaks are assigned relative to the base peak.

The peak with the highest m/z value:

- is often the molecular ion (heaviest ion), but it might be too small or undetectable in cases where it is unstable;
- will give the molecular mass and therefore the molecular formula of the compound.

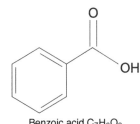

Benzoic acid $C_7H_6O_2$

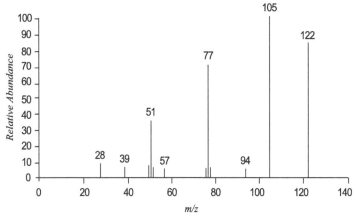

The base peak in the previous example has a m/z ratio of 105 and the molecular ion peak has a m/z ratio of 122 $[C_7H_6O_2]^+$ (molecule with one electron removed).

Mass difference	Suggested group
15	$[CH_3]^+$
17	$[OH]^+$
28	$[C=O]^+$ or $[C_2H_4]^+$
29	$[C_2H_5]^+$
31	$[CH_3O]^+$
45	$[COOH]^+$
77	$[C_6H_5]^+$

The table shows common mass differences between peaks in a mass spectrum and suggested groups that correspond to the difference.

Structure from fragmentation pattern

Compounds **A** and **B** both have the same molecular formula, C_3H_6O, but their molecular structures are different. The mass spectra of **A** and **B** are shown as follows.

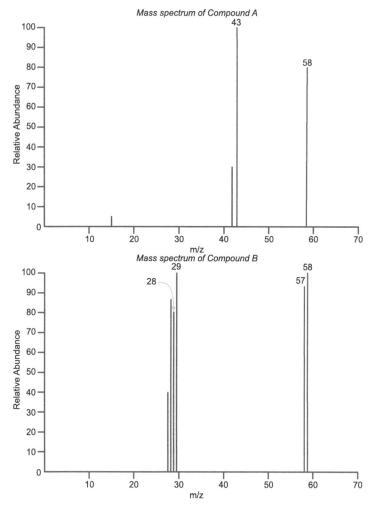

A1. For compound **A**, which group of atoms could be lost when the ion of m/z 43 forms from the ion of m/z 58?
→The mass difference from 58 to 43 is 15. This indicates that a **methyl group** is likely to be responsible for this.

A2. Suggest a formula for the ion of m/z 43 in the spectrum of compound **A**.
→The ion of m/z 43 is probably due to loss of a methyl group from the molecule, so that a fragment

of formula [C₂H₃O]⁺ would form the m/z 43 ion. A likely structure for this is [CH₃CO]⁺.

A3. Identify **A**.
→**A** is propanone, [CH₃COCH₃]⁺, which has major fragments of [CH₃CO]⁺ and [CH₃]⁺.

B1. For compound **B**, what atom or groups of atoms could be lost when

i) the molecular ion changes into the ion of m/z 57?
→A loss of 1 mass unit from 58 to 57 can only be due to loss of a **hydrogen atom**. The molecule must have a single, 'loose' H.

ii) the ion of m/z 57 changes into the ion of m/z 29?
→The mass difference from 57 to 29 is 28, probably due to loss of C=O.

B2. Suggest formulae for the ions of m/z 28,29, and 57 in the spectrum of compound **B**.
→The ion of m/z 28 is the **[C=O]⁺** fragment; addition of 1 mass unit (an **H**) to this would form a **[CHO]⁺** (aldehyde) as the m/z 29 peak.

B3. Identify **B**.
→The m/z 57 is loss of hydrogen from the molecule, leaving a fragment of formula **[C₃H₅O]⁺**. We know this contains a C=O group, so is likely to be **[C₂H₅CO]⁺**.

Questions about fragmentation Go online

Q2: Draw the structures for the six isomers of butanol - four alcohols and two ethers.

....

The mass spectra X, Y and Z as follows are for three isomers of butanol, excluding 2-methylpropan-1-ol. There is one example of a primary alcohol, a secondary alcohol and a tertiary one, but not necessarily in that order.

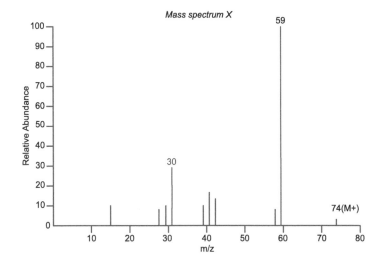

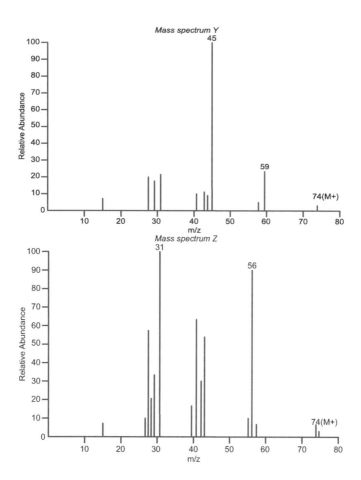

Q3: What is the m/z value for the molecular (M⁺) ion of all three butanols? Answer with one integral value.

If you would like to try to link the mass spectra with the appropriate molecular structures, have a go now, by working out the main fragments lost from the M⁺ ion. Otherwise, the remaining questions will guide you through to an answer, with the conclusions summarised at the end.

Q4: Spectrum X has a base peak at m/z 59. What fragment has been lost from the molecular ion to produce this m/z 59 ion?

....

Q5: Using the fragmentation information, which of the three butanols would be likely to produce a CH_3 fragment most readily?

Q6: Closer examination of spectrum Y shows a base peak at m/z 45. What could be the structure of this ion?

Q7: When the C-C bond adjacent to the the C-OH bond breaks in butan-1-ol what are the structures and m/z values of the fragments?

Q8: In which spectrum are these most prominent?

Q9: Given that alcohols tend to fragment at the C-C bond adjacent to the C-OH group (i.e. R-CH$_2$OH would produce R$^+$ and [CH$_2$OH]$^+$ fragments), can you explain, with reasons, which spectrum is for which isomer?

The following is the mass spectrum of naphthalene an aromatic molecule.

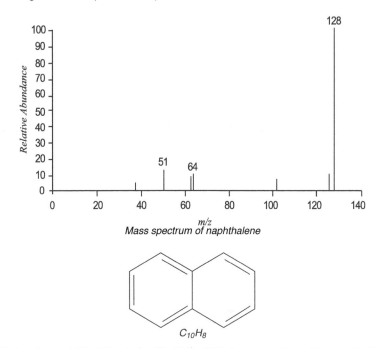

Mass spectrum of naphthalene

$C_{10}H_8$

Naphthalene has a stable molecular ion [C$_{10}$H$_8$]$^+$ which gives a very large response at m/z = 128 with minor fragmentation peaks in the rest of the spectrum. In this case the base peak is the molecular ion and naphthalene has a molecular mass of 128. The peak at m/z = 64 indicates the stability of the molecular ion being assigned to a doubly charge ion M^{2+} with m = 128 and z = 2 so that m/z = 64. This is fairly typical of aromatic compounds giving further evidence for their stability.

© HERIOT-WATT UNIVERSITY

4.3 Infrared spectroscopy

Infrared is found on the electromagnetic spectrum between microwaves and visible light (see topic on 'Electromagnetic radiation and atomic spectra'). When it is absorbed by organic compounds there is sufficient energy to cause the bonds in the molecules to vibrate but not break them.

The wavelength of the infrared radiation that is absorbed (quoted as a wavenumber cm^{-1}) by a vibrating bond depends on the type of atoms which make up the bond along with the strength of the bond (wavenumber = 1/wavelength). Light atoms that are joined by stronger bonds absorb radiation of shorter wavelengths (higher energy) in general than heavier atoms joined by weaker bonds. Infrared can be used to identify certain bonds and functional groups in organic molecules.

Simple diatomic molecules, for example halogens, have only one bond, which can vibrate only by stretching and compressing. The two halogen atoms can pull apart and then push together.

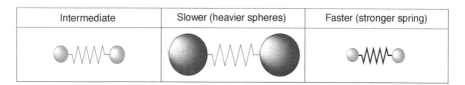

Compound	Bond enthalpy/kJ mol^{-1}
HCl	432
HBr	366
HI	298

The table shows the bond enthalpies for various hydrogen halides.

Questions on bond vibration   Go online

Q10: Which molecule has the strongest bond?

a) HCl
b) HBr
c) HI

....

Q11: Which molecule has the largest molecular mass?

a) HCl
b) HBr
c) HI

....

Q12: Which molecule will have the largest vibration frequency?

a) HCl
b) HBr
c) HI

Some organic molecules with typically more than two atoms can give rise to modes of vibration other than simple stretching of bonds. Consider water (H_2O) each of the O-H bonds will have a characteristic stretching frequency, but these can interact and stretch either symmetrically or asymmetrically, giving rise to two slightly different IR absorption frequencies for the O - H bonds in a water molecule.

Figure 4.1: Stretching modes in water molecules

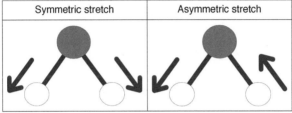

Symmetric stretch	Asymmetric stretch

Figure 4.2: Bending modes

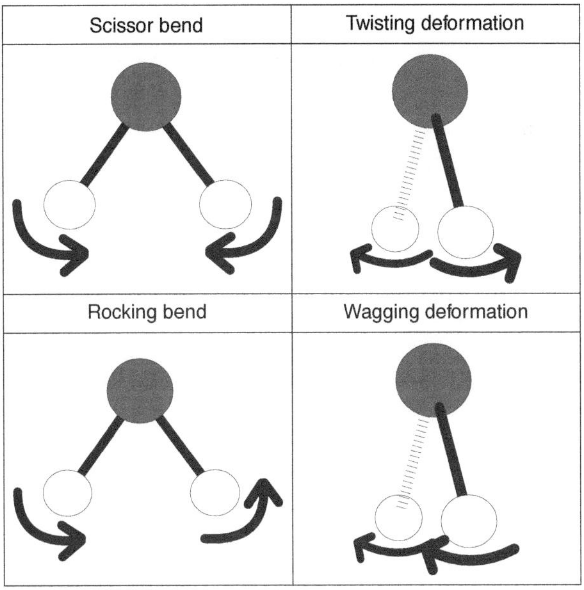

Scissor bend	Twisting deformation
Rocking bend	Wagging deformation

Infrared spectrometer

The source beam of infrared radiation gets split into two with one part of the beam passing through the sample and the other through a reference cell (may contain solvent). The monochromator grating scans the wavelengths prior to a detector which compares the intensity of the two beams. The amplified signal is plotted as % transmission or absorbance.

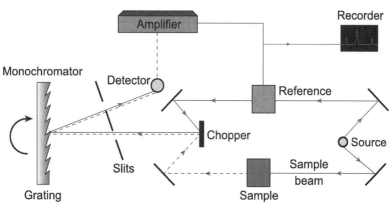

Interpreting infrared spectra

An IR spectrum of ethanol C_2H_5OH follows. An infrared correlation table can be found on page 14 of the CfE Higher and Advanced Higher Chemistry data booklet. Tables of values have been built up by observation of a large number of compounds.

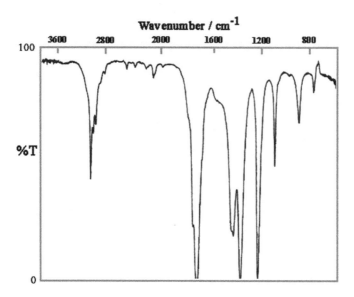

The region of an IR spectrum from 4000 to 1400 cm^{-1} contains many absorbance wavenumbers for specific bond types. Below 1400 cm^{-1} IR spectra typically have a number of absorbances not assigned to a particular bond, and this is called the fingerprint region. In fact, these relatively low energy vibrations are due to complex vibrations which are unique to that molecule. Comparison of this region for an unknown material with a set of standards run under identical conditions will allow identification.

There is a range of values for a particular group, because the value is slightly altered by the surrounding groups.

Ethanol contains a hydroxyl (OH) functional group which is assigned the wave number range of 3570 to 3200 cm^{-1} for a OH hydrogen bonded stretch. This is very broad in alcohols.

Functional group identification Go online

The main use of IR spectra, which can be obtained quickly and cheaply, is to identify the presence of functional groups and the carbon backbone type in unknown organic compounds.

Use the correlation table on page 14 of the data booklet to answer the following questions regarding the IR spectra labelled X, Y and Z.

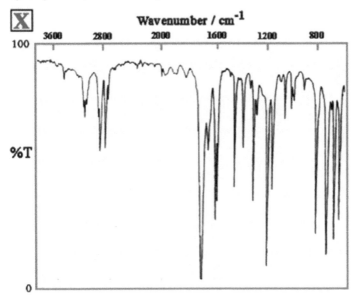

© HERIOT-WATT UNIVERSITY

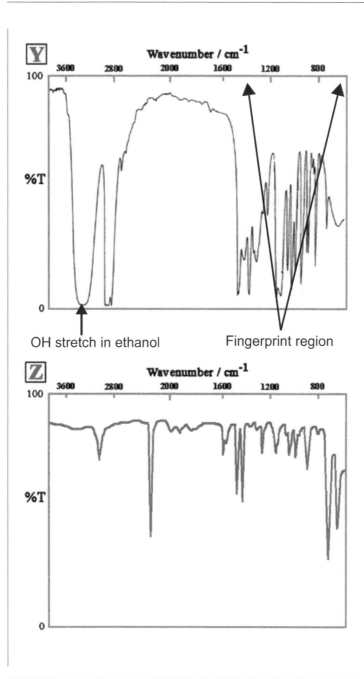

Q13: Which compound contains an alcohol group?

a) X
b) Y
c) Z

....

Q14: Which compound is **not** aromatic?

a) X
b) Y
c) Z

....

Q15: Which class of compound is present in X? Hint: take note of the peaks in the region of 1700cm^{-1} and 2800cm^{-1}.

....

Q16: Which of the following could be Z?

a) Benzyl alcohol ($C_6H_5CH_2OH$)
b) Butan-1-ol (C_4H_9OH)
c) Benzonitrile (C_6H_5CN)
d) Acetonitrile (CH_3CN)

4.4 Proton nuclear magnetic resonance (NMR) spectroscopy

Proton nuclear magnetic resonance spectroscopy (proton NMR) can give information about the different environments of hydrogen atoms in an organic molecule, and about how many hydrogen atoms there are in each of these environments.

Hydrogen nuclei (protons) behave like tiny magnets as a result of them spinning in a clockwise direction or anticlockwise direction. When they are placed between the poles of a very powerful magnet some of the protons align with the field of the magnet and some align themselves against the field of the magnet. A superconducting magnet cooled with liquid helium and liquid nitrogen is used.

The protons that are aligned with the direction of the magnetic field have a slightly lower energy than those aligned against and the energy difference between the two corresponds to the radiofrequency region of the electromagnetic spectrum (60 MHz to 1000MHz). When the protons are exposed to radio waves, the absorption of energy promotes those in the lower energy states to the higher energy states. This effectively flips those aligned with the magnetic field to being aligned against it. The protons then fall back to the lower energy state and the same radio frequency that was absorbed is emitted which can be measured with a radio receiver.

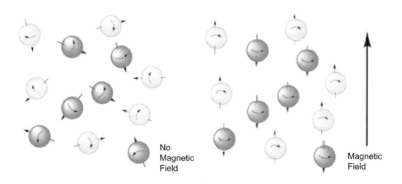

The darker spheres have lower energies than the lighter spheres and are aligned with the direction of the magnetic field in the second diagram.

Diagram of a typical NMR spectrometer

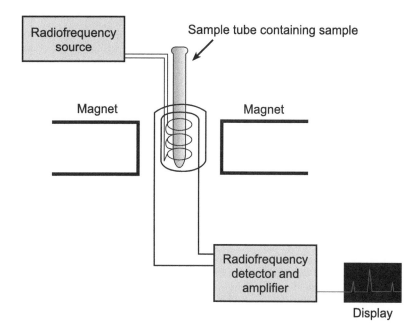

The sample being analysed is dissolved in $CDCl_3$ or CD_3COCD_3, 'deuterated' solvents with hydrogens replaced by deuterium (isotope of H which is written as D or ^2H). This is to avoid swamping the NMR spectrum of the sample with a proton signal from the solvent. A referencing substance called tetramethylsilane (TMS) is used, against which all other absorptions due to other proton environments are measured. It has only one proton environment and is assigned a value of 0 ppm (parts per million). The difference between the protons in TMS and other proton environments is called the chemical shift measured in ppm and given the symbol δ. Values for chemical shifts for different proton environments are given on page 16 of the data booklet.

Interpreting NMR spectra

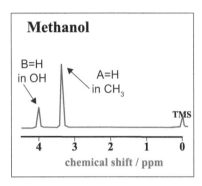

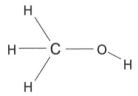

The ^1H NMR spectrum of methanol (CH_3OH) is shown in the diagram. Methanol has 2 different proton environments the H in the CH_3 group and the H in the OH group. The spectrum shows two peaks which correspond to the two different proton environments. The area under each peak is proportional to the number of protons in that particular environment. The area under peaks A and B are in the ratio 3:1 corresponding to the three protons in the CH_3 group and the one proton in the OH group. If you look at the table of chemical shifts in the data book page 16 you will see the chemical shift for R-OH is 1.0-5.0 and the chemical shift for CH_3O is 3.5-3.9 which makes the peak for the protons in CH_3 A and the peak for the proton in OH B.

Determining structure using NMR

Go online

Figure 4.3: NMR spectrum of hydrocarbon W

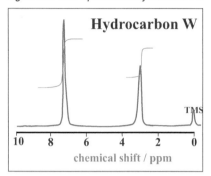

Q17: From the spectrum, how many different environments are there for hydrogen atoms?

....

Q18: What is the ratio of areas for these peaks?

....

Q19: From the correlation table (page 16 of the data booklet), can you identify the type of hydrogen in the larger peak?

a) RCH_3
b) RCH_2R
c) $ArCH_3$
d) Ar**H**

....

Q20: What type of hydrogens are in the smaller peak?

a) RCH_3
b) RCH_2R
c) $ArCH_3$
d) Ar**H**

You might find it useful to summarise the information in a table.

Table 4.1: NMR table

	Peak 1	Peak 2
δ (ppm)	2.3	7.4
Type of H	$ArCH_3$	Ar**H**
Number of H atoms	3	5
Group	CH_3	C_6H_5

Q21: What is the hydrocarbon?

The next three questions refer to N,N-diethylphenylamine ($C_{10}H_{15}N$), shown in Figure 4.4.

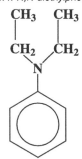

Figure 4.4: N,N-diethylphenylamine

Q22: How many different types of H environment are there in the molecule?

a) 1
b) 2
c) 3
d) 5

....

Q23: What is the ratio of H atoms in these groups?

a) 2:1
b) 4:5:6
c) 1:2:3
d) 3:2:5

....

Q24: Can you predict the chemical shift (δ) ranges for this compound?

....

Q25: Sketch the NMR spectrum for N,N-diethylphenylamine.

342 UNIT 4. ORGANIC CHEMISTRY AND INSTRUMENTAL ANALYSIS

High resolution NMR

In a high resolution spectrum, you find that many of what looked like single peaks in the low resolution spectrum are split into clusters of peaks. The amount of splitting tells you about the number of hydrogens attached to the carbon atom or atoms next door to the one you are currently interested in.

The number of sub-peaks in a cluster is one more than the number of hydrogens attached to the next door carbon(s) (n+1 rule where n is the number of hydrogens attached to the neighboring carbon atom).

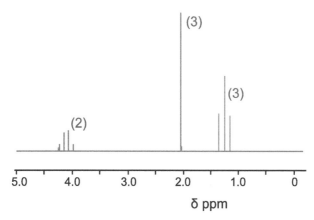

The spectrum is for the compound which has the molecular formula $C_4H_8O_2$. Treating this as a low resolution spectrum to start with, there are three clusters of peaks and so three different environments for the hydrogens. The hydrogens in those three environments are in the ratio 2:3:3. Since there are 8 hydrogens altogether, this represents a CH_2 group and two CH_3 groups.

The CH_2 group at about 4.1 ppm is a quartet. That tells you that it is next door to a carbon with three hydrogens attached - a CH_3 group.

The CH_3 group at about 1.3 ppm is a triplet. That must be next door to a CH_2 group.

This combination of these two clusters of peaks - one a quartet and the other a triplet - is typical of an ethyl group, CH_3CH_2. It is very common.

Finally, the CH_3 group at about 2.0 ppm is a singlet. That means that the carbon next door doesn't have any hydrogens attached. You would also use chemical shift data to help to identify the environment each group was in, and eventually you would come up with:

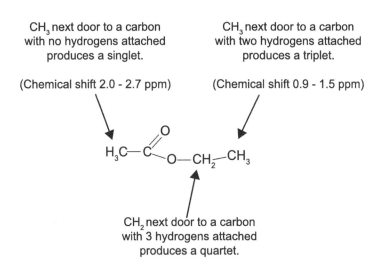

(Chemical shift 3.5 - 3.9 ppm)

The numbers represent the number of hydrogen atoms within that environment. The splitting gives us information on the number of hydrogens on neighbouring carbon atoms. Chemical shifts have been taken from the data book page 16.

Hydrogen atoms attached to the same carbon atom are said to be equivalent. Equivalent hydrogen atoms have no effect on each other - so that one hydrogen atom in a CH_2 group doesn't cause any splitting in the spectrum of the other one. But hydrogen atoms on neighbouring carbon atoms can also be equivalent if they are in exactly the same environment. For example:

$Cl-CH_2-CH_2-Cl$

These four hydrogens are all exactly equivalent. You would get a single peak with no splitting at all.

UNIT 4. ORGANIC CHEMISTRY AND INSTRUMENTAL ANALYSIS

NMR - Practice

For the high resolution ^{1}H NMR data as follows, work out the structure of the molecules concerned. You will find a short table of useful chemical shifts at the end of the questions.

Q26: A molecule with the molecular formula C_4H_8O:

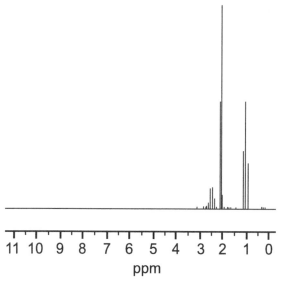

Chemical shift (ppm)	2.449	2.139	1.058
Ratio of area under the peaks	2	3	3
Splitting	quartet	singlet	triplet

....

Q27: A molecule with the molecular formula $C_4H_8O_2$:

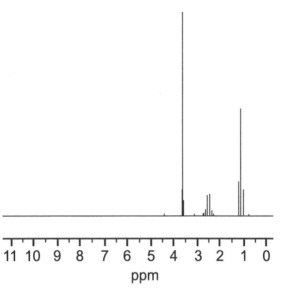

Chemical shift (ppm)	3.674	2.324	1.148
Ratio of area under the peaks	3	2	3
Splitting	singlet	quartet	triplet

....

Q28: Another molecule with the molecular formula $C_4H_8O_2$:

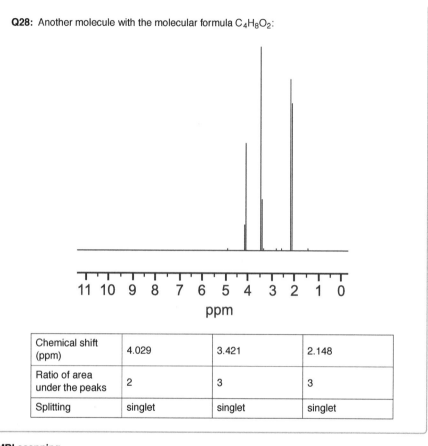

Chemical shift (ppm)	4.029	3.421	2.148
Ratio of area under the peaks	2	3	3
Splitting	singlet	singlet	singlet

MRI scanning

Our bodies consist largely of water, which exists in a large number of different environments in each tissue. Just as hydrogen atoms in molecules are shielded to different extents depending on the surrounding atoms, the protons in water in tissues will experience slight differences in a strong external magnetic field, and so will absorb slightly different radiofrequency radiation.

The part of the body being investigated is moved into a strong magnetic field. By using computers to process the absorption data, a series of images of the different water molecule environments, is built up into a picture of the body's tissues. It is assumed that MRI scanning is harmless to health, unlike other imaging processes (such as CAT scans) which involve low doses of ionising X-rays.

MRI scanning is particularly good for brain tissue where there is a large amount of fatty lipids which provides a different environment for water compared with the other tissues.

The MRI image opposite shows fluid collection in the region that separates the brain from the skull. This is a blood clot which applies pressure to the brain and is very dangerous, perhaps even fatal.

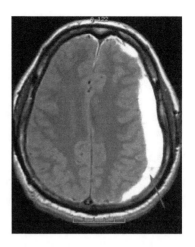

4.5 Absorption of visible light

Most organic molecules appear colourless because the energy difference between the highest occupied molecular orbital (**HOMO**) and the lowest unoccupied molecular orbital (**LUMO**) is relatively large resulting in the absorption of light in the ultraviolet region of the spectrum. Organic molecules containing only sigma bonds are colourless.

Organic molecules that are coloured contain delocalised electrons spread over a few atoms and they are known as **conjugated** systems. We have looked at conjugation in benzene. For bonds to be conjugated in long carbon chains alternating double and single bonds must be present. An example of this is seen in the structure of Vitamin A.

Conjugated system in retinol (vitamin A)

The conjugated system in Vitamin A is a long chain of σ and π bonds. The molecular orbital contains delocalised electrons which span across the length of the conjugated system. The greater the number of atoms spanned by the delocalised electrons, the smaller the energy gap will be between the delocalised orbital and the next unoccupied orbital. Exciting the delocalised electrons will therefore require less energy. If this energy corresponds to a wavelength that falls within the visible part of the electromagnetic spectrum it will result in the compounds appearing coloured.

Ninhydrin (2,2-Dihydroxyindane-1,3-dione) is a chemical used to detect amino acids. When it reacts with amino acids it produces a highly conjugated product which absorbs light in the visible region and an intense purple colour is seen.

Ninhydrin

Azo dye can be synthesised from aminobenzene (aniline), sodium nitrite and 2-naphthol at low temperatures which then can be used to dye a piece of cotton. Synthetic indigo can also be prepared using a microscale method (see RSC references on methods to carry this out).

4.6 Chromophores

A **chromophore** is a group of atoms within a molecule that is responsible for its colour. Coloured compounds arise because visible light is absorbed by the electrons in the chromophore, which are then promoted to a higher energy molecular orbital.

The red box shows the chromophore

Vitamin A (retinol) has a conjugated system that spreads over 5 carbon to carbon double bonds and appears yellow.

β-carotene found in sweet potatoes, carrots and apricots has a conjugated system that spreads over 11 carbon to carbon bonds and appears orange.

TOPIC 4. EXPERIMENTAL DETERMINATION OF STRUCTURE

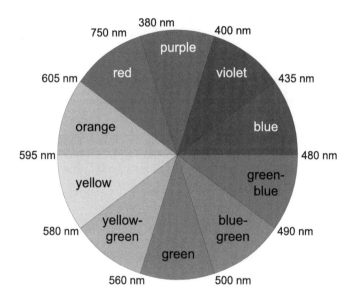

Lycopene is found in watermelon, pink grapefruit and tomatoes and has as conjugated system that spreads over 13 carbon to carbon bonds and appears red.

The colours we observe are not absorbed by the molecule. If the chromophore absorbs light of one colour, then the complementary colour is observed. This means that the Vitamin A will absorb blue light to appear yellow, β-carotene will absorb green-blue light to appear orange and lycopene will absorb blue-green light to appear red (see the following colour wheel).

Colour wheel

The energy associated with a particular colour can be calculated using the equation $E = Lhc/\lambda$ (see 'Electromagnetic radiation and atomic spectra').

4.7 Summary

Summary

You should now be able to:

- explain that elemental microanalysis can be used to determine the masses of C, H, O, S and N in a sample of an organic compound in order to determine its empirical formula;
- work out the empirical formula from data given;
- explain that mass spectrometry can be used to determine the accurate molecular mass and structural features of an organic compound;
- explain infra-red spectroscopy can be used to identify certain functional groups in an organic compound and work out which compound is responsible for a spectra by identifying which functional groups are responsible for peaks;
- proton nuclear magnetic resonance spectroscopy (proton NMR) can give information about the different environments of hydrogen atoms in an organic molecule, and about how many hydrogen atoms there are in each of these environments;
- be able again to identify which compound is responsible for a spectra and be able to draw low resolution NMR spectra;
- explain how absorption of visible light by organic molecules occurs;
- state that the chromophore is the group of atoms within a molecule which is responsible for the absorption of light in the visible region of the spectrum.

4.8 Resources

- Royal Society of Chemistry, Spectroscopy in a Suitcase (http://rsc.li/1VVGlQG)
- LearnChemistry, Introduction to Spectroscopy (http://rsc.li/2aiTp0z)
- LearnChemistry, SpectraSchool (http://rsc.li/2adQ8lo)
- Chemguide, What is Nuclear Magnetic Resonance (NMR)? (http://bit.ly/2acLnnt)
- UV-Visible Absorption Spectra (http://bit.ly/2aMflzk)
- LearnChemistry, Microscale Chemistry - The microscale synthesis of azo dyes (http://rsc.li/2a6vZYl)
- LearnChemistry, Microscale Chemistry - The microscale synthesis of indigo dye (http://rsc.li/2ay2pBa)
- Additive colour model (Flash animation) (http://bit.ly/1lY1zBt)

4.9 End of topic test

End of Topic 4 test Go online

Q29: A mass spectrum contains a line at m/z 28.0312. Given the following accurate atomic masses, what could the formula for the ion be?

H	1.0078
C	12.0000
N	14.0031
O	15.9949

a) $[NO]^+$
b) $[C_2H_4]^+$
c) $[CO]^+$
d) $[N_2]^+$

Look at the mass spectrum of butyl ethanoate.

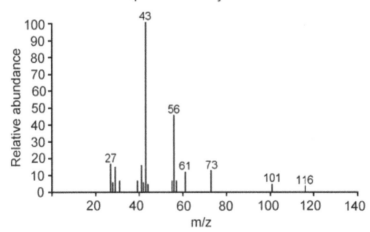

Q30: Draw the structural formula for butyl ethanoate.

....

Q31: The spectrum shows a line at m/z 116. What name is normally given to the species represented by this line?

....

Q32: The abundance of the m/z 116 ion is small because the parent molecular ion is:

a) unstable and easily fragments to form daughter ions.
b) heavy and easily fragments to form lighter ions.
c) large and easily fragments to form smaller ions.
d) charged and easily fragments to form neutral ions.

....

Q33: Write the formula of an ion which would give the line at **m/z 43**.

....

Q34: Write the formula of an ion which would give the line at **m/z 73**.

....

Q35: From the spectrum, give the m/z value of the ion which you would expect to be deflected most in the magnetic field of the mass spectrometer.

....

Q36: Ethyl butanoate is isomeric with butyl ethanoate but gives a different mass spectrum. By comparing the structures of the esters, which fragment could **not** be present in both spectra?

a) $[C_4H_9]^+$ at m/z 57
b) M^+ at m/z 116
c) $[C_2H_5]^+$ at m/z 29
d) $[C_3H_7]^+$ at m/z 43

....

Q37: Which **two** ions from the following options could produce a peak with m/z 29 in a mass spectrum?

a) $[CH_3]^+$
b) $[CHO]^+$
c) C_6H_6
d) C_2H_2
e) $[C_2H_5]^+$
f) $[CH_2OH]^+$

....

Q38: Which **two** substances from the following options have an empirical formula CH?

a) $[CH_3]^+$
b) $[CHO]^+$
c) C_6H_6
d) C_2H_2
e) $[C_2H_5]^+$
f) $[CH_2OH]^+$

TOPIC 4. EXPERIMENTAL DETERMINATION OF STRUCTURE

Q39: In mass spectra where a molecular ion loses 15 mass units to produce an (M-15)$^+$ ion, what group is most likely to have been lost?

a) $[CH_3]^+$
b) $[CHO]^+$
c) C_6H_6
d) C_2H_2
e) $[C_2H_5]^+$
f) $[CH_2OH]^+$

Q40: Which **two** fragment ions could be produced when the C - C bond in ethanol breaks?

a) $[CH_3]^+$
b) $[CHO]^+$
c) C_6H_6
d) C_2H_2
e) $[C_2H_5]^+$
f) $[CH_2OH]^+$

Q41: In proton nuclear magnetic resonance spectroscopy, which of these would you expect to be used as an internal reference standard?

a) $CDCl_3$
b) Deuterium
c) $Si(CH_3)_4$
d) Hydrogen

Q42: Which of the following compounds is most likely to show an infrared absorption at 1725 cm^{-1}?

a) 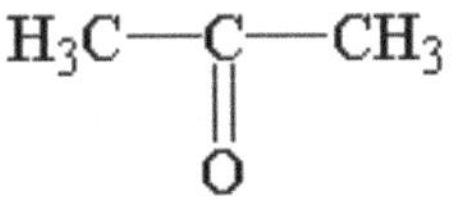 $H_3C-\underset{\underset{O}{\|}}{C}-CH_3$

b) $HOCH_2CH=CH_2$

c) 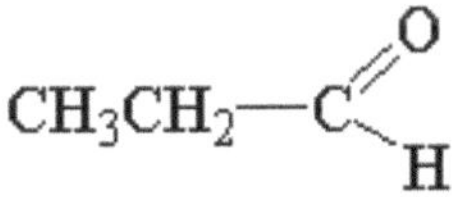 $CH_3CH_2-C\underset{H}{\overset{O}{\diagup\!\!\!\diagdown}}$

d) 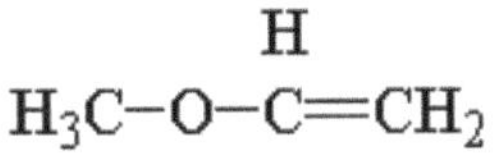 $H_3C-O-\underset{H}{\overset{}{C}}=CH_2$

Q43: How many peaks would you expect in the ^{1}H NMR spectrum of cyclohexane (C_6H_{12})?

a) 1
b) 2
c) 6
d) 12

....

Q44: Consider the ^{1}H NMR spectrum of the ether methoxyethane ($CH_3OCH_2CH_3$). How many peaks would there be?

....

Q45: What is the ratio of the areas of the peaks? If you think there are two peaks in a ratio 1:2, enter your answer in that form.

Two aliphatic compounds A and B, which contain carbon, hydrogen and oxygen only, are isomers. They can both be oxidised as follows:

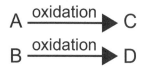

The following table shows the wave numbers of the main absorptions in the infrared spectra caused by the functional groups in compounds A to D, between 1500 and 4000 cm^{-1} (absorptions caused by C - H bonds in alkyl components have been omitted):

Compound	Wavenumbers		
A	3300		
B	3350		
C		2750	1730
D			1700

In answering the following questions, you are advised to consult the data on **page 14 of the Data Booklet**.

Q46: Examine the preceding table and identify the compound present in both A and B.

....

Q47: Which type of compound is compound C?

....

Q48: Which type of compound is compound D?

....

TOPIC 4. EXPERIMENTAL DETERMINATION OF STRUCTURE

Q49: Compound C will undergo further oxidation to produce compound E. Estimate the wavenumber ranges of the main infrared absorptions caused by the functional groups in compound E (only those between 1500 and 4000 cm^{-1}).

An unpleasant smelling liquid has spilled from a tank which has its label obscured. A sample has an NMR spectrum with only one peak.

Q50: Which **two** substances from the following list could the liquid be?

a) 1,2-dichloroethane, CH_2ClCH_2Cl
b) Propylamine, $C_3H_7NH_2$
c) Ethanenitrile, CH_3CN
d) Ethanal, CH_3CHO

....

Q51: The IR spectrum has strong absorbances at 2930, 2245 and 1465 cm^{-1}. The material is most likely to be:

a) 1,2-dichloroethane, CH_2ClCH_2Cl
b) Propylamine, $C_3H_7NH_2$
c) Ethanenitrile, CH_3CN
d) Ethanal, CH_3CHO

....

Q52: In what range would you expect the NMR chemical shift of the ^1H signal for this compound to be?

....

Q53: Anthocyanins are responsible for the colours of many plants and fruits. One such anthocyanin is pelargonidin which produces an orange colour. The structure of this is shown as follows:

Using your knowledge of chemistry explain why an orange colour is produced. *(3 marks)*

Unit 4 Topic 5

Pharmaceutical chemistry

Contents
    5.1 Effect of drugs on the body 358
    5.2 How drugs work 361
    5.3 Pharmacological activity 361
    5.4 Summary 362
    5.5 Resources 362
    5.6 End of topic test 363

Learning objective

By the end of this topic, you should be able to state that:

- drugs are substances which alter the biochemical processes in the body;
- many drugs can be classified as agonists or as antagonists at receptors, according to whether they enhance or block the body's natural responses;
- drugs bind to receptors in the body so that each drug has a structural fragment which confers the pharmacological activity.

5.1 Effect of drugs on the body

Drugs are substances which alter the biochemical processes in the body. Drugs which have beneficial effects are used in medicines. A medicine usually contains the drug plus other ingredients.

Paracetamol

Paracetamol is a mild analgesic (pain reliever) and is used to alleviate pain from headaches and all other minor aches and pains. It is also a major ingredient in many cold and flu remedies. While generally safe for use at recommended doses, even small overdoses can be fatal. Compared to other over-the-counter pain relievers, paracetamol is significantly more toxic in overdose potentially leading to liver failure and death, but may be less toxic when used persistently at the recommended dosage.

Classification of drugs

Most medicines can be classed as **antagonists** or **agonists** according to the response they trigger (enhance or block) when bound to a receptor site. An agonist will mimic the body's naturally active molecule so when bonded to the receptor site it produces the same response as the body's own molecule would do. Antagonists bind strongly to the receptor site, blocking the body's natural molecule from binding and preventing the triggering of the natural response.

An antagonist therefore is a drug which binds to a receptor without stimulating cell activity and prevents any other substances from occupying that receptor. This class of medicine is useful if there is a surplus of natural messengers or where one wants to block a particular message. For example, propranolol is an antagonist which blocks the receptors in the heart that are stimulated by adrenaline. It is called a β-blocker and is used to relieve high blood pressure.

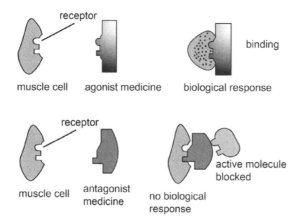

An example of an agonist medicine is salbutamol (as shown in the following diagram) which is used in the treatment of asthma. Asthma attacks are caused when the bronchioles narrow becoming blocked with mucus. The body responds by releasing adrenaline which binds to receptors triggering the dilation of the bronchioles. Unfortunately it also triggers an increase in the heart rate and blood pressure which could lead to a heart attack. Salbutamol binds more strongly to the receptor sites than adrenaline triggering the widening of the bronchioles without an increase in heart rate.

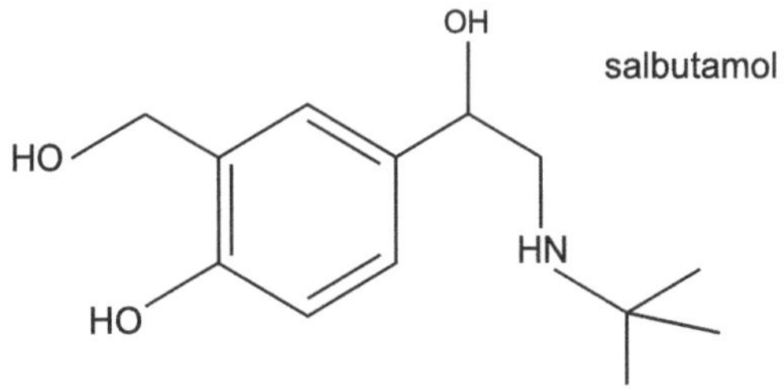

Another example of an antagonist medicine is propranolol (as shown in the following diagram) which is used to treat high blood pressure.

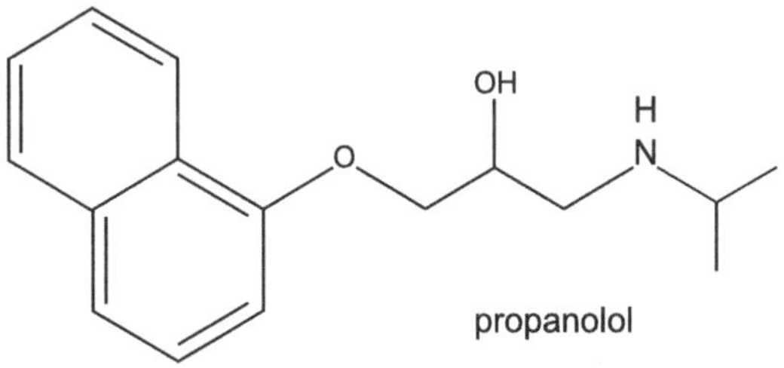

Anandamide (bliss molecule) is a recently discovered messenger molecule that plays a role in pain, depression, appetite, memory, and fertility. Anandamide's discovery may lead to the development of an entirely new family of therapeutic drugs. The structure is very similar to that of tetrahydro-cannabinol (THC) the active constituent of cannabis and marijuana.

For many years scientists wondered why compounds such as morphine, which were derived from plants, should have a biological effect on humans. They reasoned that there must be a receptor in the brain that morphine could bind to. Scientists suggested that the brain had its own morphine-like molecule, and the receptors were meant for them (morphine just happened to look like the brain's molecule and so had a similar effect). These morphine-like molecules were eventually discovered and called enkephalins, the body's natural painkillers. In 1988, specific receptors were discovered in the brain for THC and scientists started to look for the brain's natural analogue of THC which was isolated in 1992 and called anandamide.

Anandamide has a long hydrocarbon tail which makes it soluble in fat and allows it to easily slip across the blood-brain barrier. Its 3-dimensional shape strongly resembles that of THC. However, THC is a relatively robust molecule, whereas anandamide is fragile and breaks down rapidly in the body. That is why anandamide doesn't produce a continual 'natural high'.

Anandamide is synthesised enzymatically in the areas of the brain that are important in memory, thought processes and control of movement. Research suggests that anandamide plays a part in the making and breaking of short-term connections between nerve cells related to learning and memory. Animal studies suggest that too much anandamide induces forgetfulness suggesting that if substances could be developed that keep anandamide from binding to its receptor, these could be used to treat memory loss or even to enhance existing memory. Anandamide also acts as a messenger between the embryo and the uterus during implantation into the uterine wall.

Anandamide occurs in minute quantities in a number of organisms, from sea urchin roe, pig brains and mice livers. Three compounds that strongly resemble anandamide were found in dark chocolate which is why we may get pleasure from eating chocolate.

Extended information: Lipinski's rule of five

Lipinski's rule of five is a rule of thumb to evaluate drug likeness (qualitative concept used in drug design) or determine if a chemical compound with a certain pharmacological or biological activity has properties that would make it a likely orally active drug in humans. The rule was formulated by Christopher A. Lipinski in 1997, based on the observation that most orally administered drugs are relatively small and moderately lipophilic (fat loving) molecules. Orally active drugs must not break any more than one of the following criteria.

> Molecular mass less than 500 amu.
>
> Not more than 5 hydrogen bond donors.
>
> Not more than 10 hydrogen bond acceptors.
>
> An octanol/water partition coefficient log p not greater than 5.

5.2 How drugs work

Most medicines work by binding to receptors within the body which are normally protein molecules. Some of these protein receptor molecules are embedded in the membrane that surrounds cells while others are located within the cell. Those located inside cells are enzymes, globular proteins which act as catalysts in the body. These receptors are referred to as catalytic receptors and their binding sites are often called active sites. Molecules which bind to these active sites are normally called substrate molecules and the catalytic receptors catalyse a reaction on the substrate molecules. It is critical that the shape of the binding molecule complements the shape of the receptor site. Once the molecule fits into the receptor site several forces including hydrogen bonds, London dispersion and ionic bonds hold it in place to prevent it floating away.

Protein receptors that are embedded in membranes have hollows or clefts in their surface into which small biologically active molecules can fit and bind. The binding of the active molecule triggers a response.

5.3 Pharmacological activity

Analgesic medicines (pain relief) including morphine and codeine have a common structural fragment (known as a pharmacophore). These drugs are highly addictive and are strictly controlled legally.

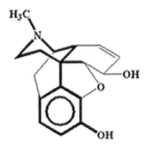
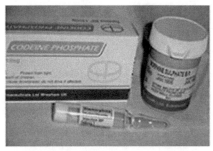

The part of the molecule shown in bold is the structural fragment common to many analgesic medicines. When the shape, dimensions and functional groups present have been identified, this can be cropped, added to and manipulated by chemists to produce compounds which are still analgesic but less addictive. The shape and dimensions of this structural fragment complement the receptor usually through functional groups that are positioned by the shape and size of the structures to allow interaction and binding of the medicine to the receptor.

benzomorphan **etorphine**

Etorphine (a synthetic opiate) is almost 100 times as potent as morphine and is used in veterinary medicine to immobilize large animals.

5.4 Summary

Summary

You should now be able to state that:

- drugs are substances which alter the biochemical processes in the body;
- many drugs can be classified as agonists or as antagonists at receptors, according to whether they enhance or block the body's natural responses;
- drugs bind to receptors in the body so that each drug has a structural fragment which confers the pharmacological activity.

5.5 Resources

- Anandamide: The molecule of extreme pleasure (http://bit.ly/2acWWe9)
- Anandamide (http://bit.ly/1EvLXX0)
- View 3D Molecular Structures (https://www.pymol.org/view)

5.6 End of topic test

End of Topic 5 test Go online

Q1: Calculate the % m/v concentration of sodium ions, if 5 g of sodium chloride is dissolved in 500 cm^3 of water.

....

Q2: Calculate the volume of ethanoic acid in a 250 cm^3 bottle of vinegar, if the % v/v concentration is 3.5%.

....

Q3: A strip of toothpaste weighing 0.75 g was found to contain 1.125 mg of fluoride. Calculate the concentration of fluoride in the toothpaste in ppm.

....

Q4: Calculate the mass of sodium fluoride in mg which would need to be added to 10 g of toothpaste to give this concentration.

Look at this diagram of four chemical compounds.

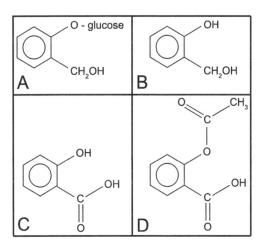

Q5: Which of the diagrams represents aspirin?

....

Q6: Which reaction type takes place when D is converted to C?

a) Esterification
b) Reduction
c) Hydrolysis
d) Oxidation

Q7: Which reaction type takes place when C is converted to B?

a) Esterification
b) Reduction
c) Hydrolysis
d) Oxidation

Q8: Which is the molecular formula for the compound in box D?

a) $C_9H_8O_4$
b) $C_7H_4O_4$
c) $C_9H_4O_4$
d) $C_9H_{10}O_4$

These structures show isoprenaline and salbutamol. Both can be used in the treatment of asthma.

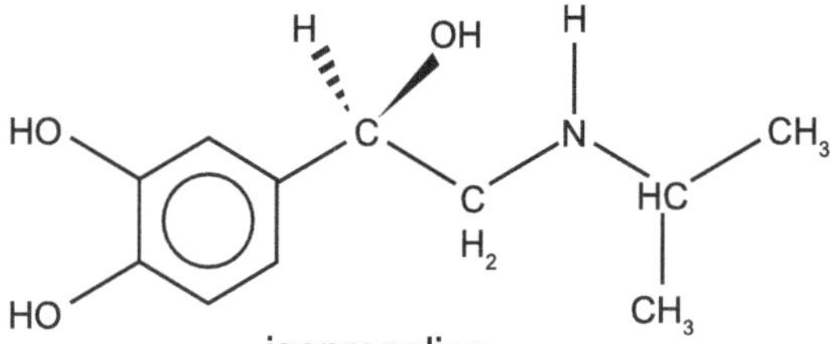

isoprenaline

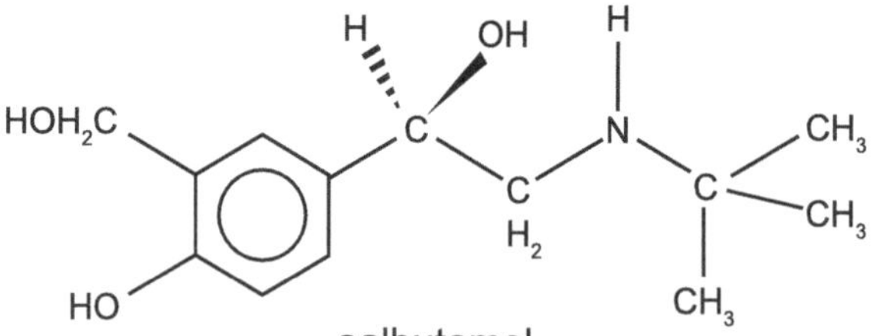

salbutamol

Q9: Which of these functional groups is not present in either molecule?

a) Amine
b) Alkene
c) Phenyl
d) Hydroxyl

TOPIC 5. PHARMACEUTICAL CHEMISTRY

Q10: Which of these classes of alcohol is present in salbutamol but not in isoprenaline?

a) Primary
b) Secondary
c) Tertiary
d) Aromatic

This structure is part of both adrenaline and salbutamol and is the fragment upon which their activity depends.

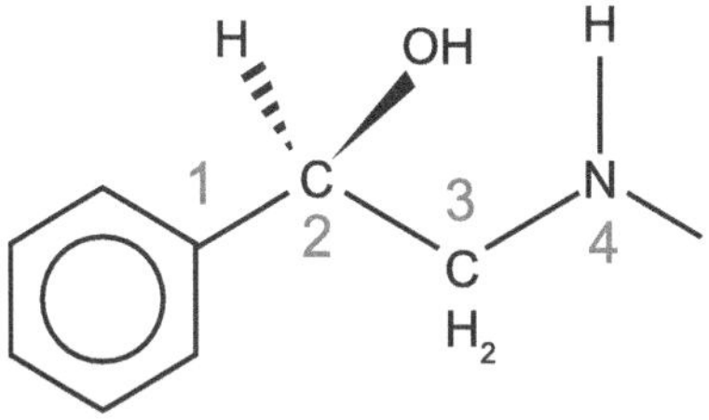

Q11: Salbutamol is a chiral molecule. Which atom indicates the chiral centre?

Look at these three analgesics. Aspirin and paracetamol act on pain by occupying the enzyme site needed to make the prostaglandins which are produced in response to injury. Phenacetin is metabolised in the body to paracetamol.

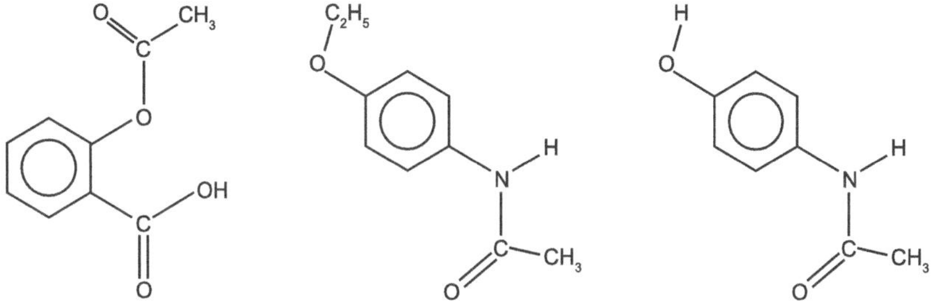

A. aspirin   B. phenacetin   C. paracetamol

Q12: Which of these is an ether?

....

Q13: Which of these is a phenol?

....

Q14: Which of these is a carboxylic acid?

....

© HERIOT-WATT UNIVERSITY

Q15: What name is used to describe a medicine which acts to block the body's natural responses?

....

Q16: The metabolism of phenacetin also produces ethanal. Which additional element would have to be added to phenacetin to give paracetamol?

....

Q17: Salbutamol is a **chiral** molecule. One of the **enantiomers** is 68 times more active than the other. Explain fully why one form of salbutamol is much more active than the other. Use the emboldened words in your answer.

Unit 4 Topic 6

Organic chemistry and instrumental analysis test

Organic chemistry and instrumental analysis test

Go online

Q1: Propene molecules contain:

a) sp^2 hybridised carbon atoms but no sp^3 hybridised carbon atoms.
b) sigma bonds but no pi bonds.
c) sp^2 hybridised carbon atoms and sp^3 hybridised carbon atoms.
d) pi bonds but no sigma bonds.

....

Q2: Which is the correct description of the numbers of sigma and pi bonds in buta-1,3-diene in the following table?

	Number of sigma bonds	Number of pi bonds
A	7	4
B	9	2
C	7	2
D	2	7

....

Q3: The electronic structure of a carbon atom ($1s^2$ $2s^2$ $2p^2$) suggests that carbon has two unpaired electrons, however, in alkanes, all the carbon atoms form four bonds. Explain this observation.

....

Q4: Which one of these is the skeletal structure of C_4H_{10}?

A

B

C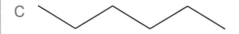

D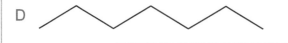

....

Q5: What is the molecular formula of the following skeletal structure? Also name the compound.

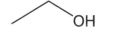

....

Q6: Draw the full structure of the following skeletal structure and name the compound.

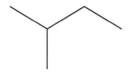

....

Q7: Two optical isomers were made. The isomers could be distinguished using what test?

a) Solubility
b) Effect on plane of polarised light
c) Melting point and boiling points
d) Ease of dehydration

....

Q8: Which of the following isomeric alcohols would exhibit optical isomerism?

a) $CH_3CH_2CH_2CH_2CH_2OH$
b) $CH_3CH_2CH_2CH(OH)CH_3$
c) $CH_3CH_2CH(OH)CH_2CH_3$
d) $(CH_3)_2C(OH)CH_2CH_3$

....

Q9: One of the optical isomers of lactic acid has the following structure:

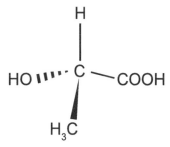

Explain why lactic acid can exphibit optical isomerism. Then draw a diagram to show the other optical isomer of lactic acid.

....

Q10: Which of the following pair of compounds react to form the ester methyl ethanoate?

a) CH₃Cl and CH₃COOH
b) CH₃OH and CH₃COCl
c) CH₃CH₂OH and HCOOH
d) CH₃COCl and CH₃Cl

Q11:

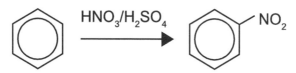

This reaction can be described as:

a) electrophilic addition.
b) electrophilic substitution.
c) nucleophilic addition.
d) nucleophilic substitution.

Lactic acid (2-hydroxypropanoic acid) can be prepared by the following 2 step process:

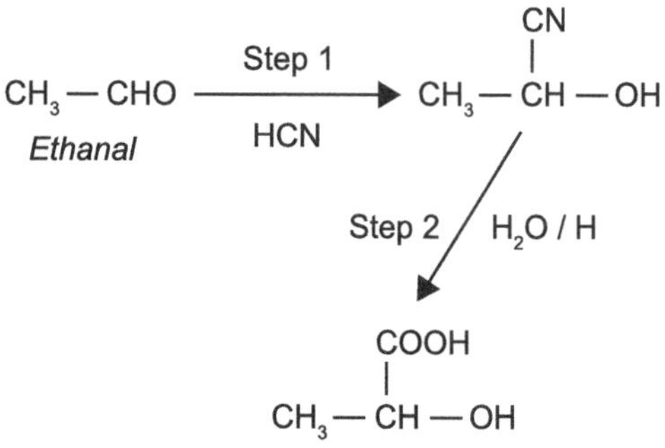

$$CH_3-CHO \xrightarrow[HCN]{Step\ 1} CH_3-\underset{\underset{OH}{|}}{\overset{\overset{CN}{|}}{C}}H \xrightarrow[H_2O/H]{Step\ 2} CH_3-\underset{\underset{OH}{|}}{\overset{\overset{COOH}{|}}{C}}H$$

Ethanal

lactic acid
(2-hydroxypropanoic acid)

Q12: Name the type of reaction taking place in Step 2.

Q13: Give the systematic name of the final product if butanone had been used instead of ethanal.

Butan-2-ol can be prepared from but-2-ene by the following route:

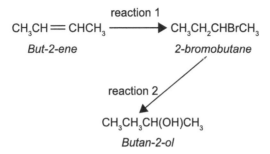

Q14: In reaction 1, but-2-ene undergoes electrophilic addition with hydrogen bromide. Draw diagrams to outline the mechanism for this reaction.

....

Q15: Name a suitable reagent to carry out reaction 2.

....

Q16: The diagram shows a simplified mass spectrum for butanone.

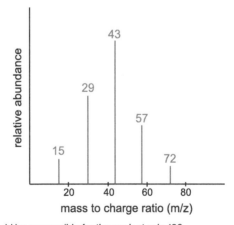

Which fragment could be responsible for the peak at m/z 43?

a) $[CH_3]^+$
b) $[C_2H_5CO]^+$
c) $[C_2H_5]^+$
d) $[CH_3CO]^+$

....

Q17: An organic compound Y has the following NMR spectrum:

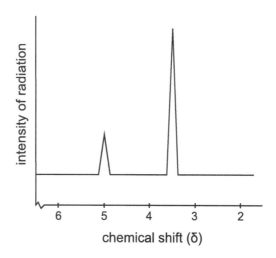

Y could be:

a) CH_3CH_3
b) CH_3NH_2
c) CH_3OH
d) CH_3OCH_3

....

Q18: What technique would be the best way of identifying the presence of a carbonyl group (C=O) in an organic compound?

a) Infra-red spectroscopy
b) Mass spectroscopy
c) Elemental microanalysis
d) Nuclear Magnetic Resonance spectroscopy

On subjecting a 0.4440 g sample of an organic compound, **X**, to elemental analysis, it was found to contain 0.2161 g of carbon and 0.0359 g of hydrogen. The only other element present in **X** was oxygen. The infra-red spectrum of **X** was also recorded.

Q19: Calculate the empirical formula of compound **X**.

....

Q20: What effect does the absorption of infra-red radiation have on the bonds in a molecule of compound **X**?

....

TOPIC 6. ORGANIC CHEMISTRY AND INSTRUMENTAL ANALYSIS TEST

Q21: The drug *ranitidine* has proved effective in healing stomach ulcers. It binds to cells in the same way as histamine and so prevents histamine triggering the release of excess hydrochloric acid which is the cause of the ulcers. In healing ulcers ranitidine is acting as:

a) agonist.
b) antagonist.
c) receptor.
d) base.

Morphine, metazocene and pethidine are analgesic drugs.

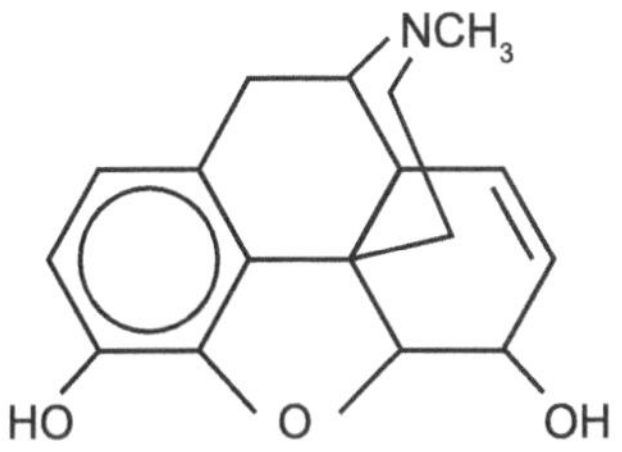

morphine

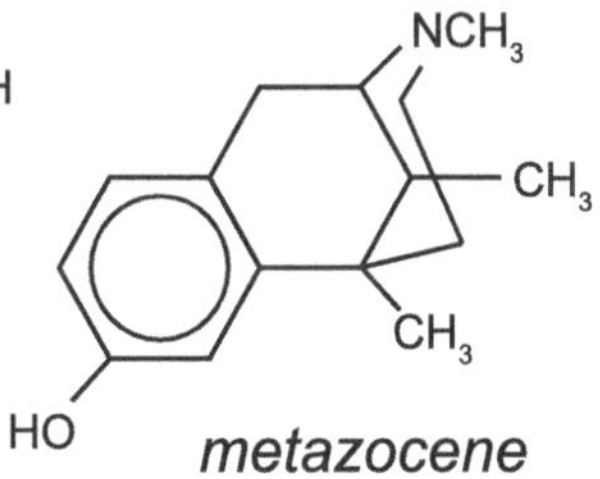

metazocene

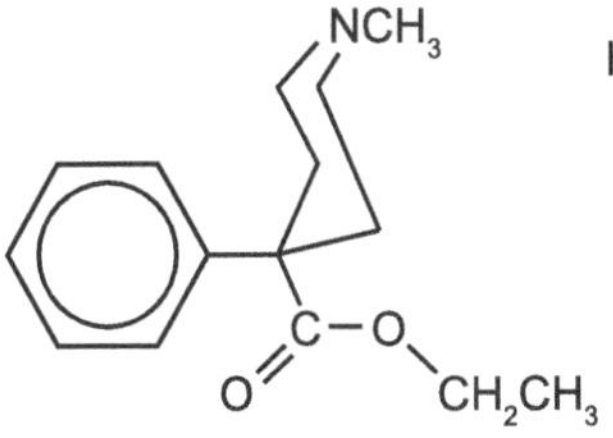

pethidine

Q22: Identify the pharmacophore in these drugs and draw its structural formula.
....

Q23: Morphine, metazocene and pethidine are agonists. What is meant by the term agonist?

Glossary

Agonist
> a drug which enhances the body's natural response or mimics the natural response of the body

Amphoteric
> a substance which can act as both an acid and a base

Antagonist
> a drug which blocks the natural response of the body

Aromatic
> aromatic compounds contain a benzene ring in their structure

Aufbau principle
> states that orbitals are filled in order of increasing energy

Avogadro's constant
> the number of constituent particles, usually atoms or molecules, that are contained in the amount of substance given by one mole ($L = 6.02 \times 10^{23}$ mol^{-1})

Buffer solution
> a solution in which the pH remains approximately constant when small amounts of acid or base are added

Carbocation
> an ion with a positively charged carbon atom

Chiral
> A chiral molecule is one which has a non-superimposable mirror image, i.e. optical isomers exist. All chiral molecules have a chiral carbon atom, i.e. a carbon atom with four different atoms or groups bonded to it

Chromophore
> the group of atoms within a molecule which is responsible for the absorption of light in the visible spectrum. Molecules containing a chromophore are coloured

Cis
> cis molecules have two of the same atom or group on the same side of a carbon-to-carbon double bond

Closed
> a closed system has no exchange of matter or energy with its surroundings

Conjugate acid
> the species left when a base accepts a proton

Conjugate base
> the species formed when an acid donates a proton

Conjugated

conjugated systems contain delocalised electrons spread over a few atoms. This is often through alternating single and double bonds

Coordination number

the number of bonds a transition metal ion forms with surrounding ligands

Dative

a bond where both electrons have come from one of the elements involved in the bond

Degenerate

a set of atomic orbitals that are of equal energy to each other are said to be degenerate

Dynamic equilibrium

a dynamic equilibrium is achieved when the rates of two opposing processes become equal, so that no net change results

Electromagnetic spectrum

the range of frequencies or wavelengths of electromagnetic radiation

Electrophile

a species which is attracted to an electron rich site; electrophiles are electron deficient and have a positive charge

Electrophilic addition

addition across a carbon-to-carbon double bond

Electrophilic substitution

substitution of a hydrogen atom on a benzene ring for an electrophile

Enantiomers

optical isomers (non-superimposable mirror images) are known as enantiomers.

Enthalpy of formation

the enthalpy change when one mole of a substance is formed from its elements in their standard states

Entropy

the degree of disorder of a system

Equivalence point

the equivalence point in a titration experiment is reached when the reaction between the titrant (added from the burette) and the titrate (in the flask) is just complete.

Free energy

the total amount of energy available to do work

Frequency

the number of wavelengths that pass a fixed point in one unit of time, usually one second

Ground state

the lowest possible electronic configuration the electrons in an atom can adopt

Heisenberg's uncertainty principle

states that it is impossible to state precisely the position and the momentum of an electron at the same instant

HOMO

highest occupied molecular orbital

Hund's rule

when degenerate orbitals are available, electrons fill each singly, keeping their spins parallel before pairing starts

Ionic product of water

$K_w = [H_3O^+][OH^-] = 1 \times 10^{-14}$ at 298 K

Ionisation energy

the first ionisation energy of an element is the energy required to remove one electron from each of one mole of atoms in the gas phase to form one mole of the positively charged ions in the gas phase

Ligand

an ion or molecule which can bind to a transition metal ion to form a complex: ligands have a negative charge or at least one lone pair of electrons

LUMO

lowest unoccupied molecular orbital

Nucleophile

a species which is attracted to a positive charge; nucleophiles are electron rich species, i.e. they have a negative charge or lone pairs of electrons

Nucleophilic substitution

a reaction in which one atom or group of atoms is substituted by a nucleophile

Parent ion

a molecular ion produced during electron bombardment in mass spectrometry. In a mass spectrum the peak with the largest m/z value represents the parent ion and therefore the molecular mass of the molecule

Pauli exclusion principle

states that no two electrons in the one atom can have the same set of four quantum numbers - as a consequence, no orbital can hold more than two electrons and the two electrons must have opposite spins

Planck's constant

the physical constant that is the quantum of action in quantum mechanics ($h = 6.63 \times 10^{-34}$ J s)

GLOSSARY

Racemic mixture
    a racemic mixture contains equal concentrations of both optical isomers. Racemic mixtures have no effect on plane polarised light

Rate determining step
    the slowest step in a reaction mechanism that governs the overall rate

Second law of thermodynamics
    the total entropy of a reaction system and its surroundings always increases for a spontaneous change

S_N1
    nucleophilic substitution, 1st order. Tertiary halogenoalkanes are likely to take part in S_N1 reactions, although the kinetics of these reactions can only be determined through experimental results and not by the structure alone

S_N2
    nucleophilic substitution, 2nd order. Primary halogenoalkanes are likely to take part in S_N2 reactions, although the kinetics of these reactions can only be determined through experimental results and not by the structure alone

Spectrochemical series
    a list of ligands according to how strongly they split d orbitals in a transition metal complex. From largest to smallest splitting ability $CN^- > NO_2^- > NH_3 > H_2O > OH^- > F^- > Cl^- > Br^- > I^-$

Standard conditions
    298 K (25°C) and one atmosphere pressure

Trans
    trans molecules have two of the same atom or group on opposite sides of a carbon-to-carbon double bond

Velocity
    the physical vector quantity which needs both magnitude and direction to define it, usually measured in $m\ s^{-1}$ (or m/s)

Wavelength
    the distance between adjacent crests (or troughs) and is usually measured in metres or nanometres (1 nm = 10^{-9} m)

Answers to questions and activities for Inorganic Chemistry

Topic 1: Electromagnetic radiation and atomic spectra

Obtaining wavelength from frequency (page 6)

Q1: 589 nm

Q2: 3.75×10^{-5} m

Q3: 4.29×10^{-7} m

Q4: 3.75×10^{-2} m

Q5:

$f = 2.45 \times 10^9$ Hz

and $\lambda = \dfrac{c}{f} = \dfrac{3.00 \times 10^8 \text{ m s}^{-1}}{2.45 \times 10^9 \text{ s}^{-1}}$

$\lambda = \dfrac{3.00 \times 10^8}{2.45 \times 10^9}$ m

$\lambda = 1.224 \times 10^{-1}$ m

$\lambda = 0.1224$ m

$\lambda = 12.24$ cm

Obtaining frequency from wavelength (page 7)

Q6:

$c = $ speed of light $= 3.00 \times 10^8$ m s^{-1}

$\lambda = $ wavelength $= 405 \times 10^{-9}$ m

$c = f \times \lambda$

$f = \dfrac{c}{\lambda}$

$f = \dfrac{3.00 \times 10^8 \text{ m s}^{-1}}{405 \times 10^{-9} \text{ m}}$

$f = 7.41 \times 10^{14}$ s^{-1}

Q7: c) 5.75×10^{14}

Q8: 1.58×10^{14} Hz

Q9: 1.30×10^{14} Hz

Q10: 2.00×10^{14} Hz

ANSWERS: UNIT 1 TOPIC 1

Electromagnetic radiation table (page 8)

Q11:

Quantity	Symbol	Unit	Description
speed	c	m s^{-1}	rate of travel
wavelength	λ	m	wavecrest separation
frequency	f	Hz	wave cycles per second

Using spectra to identify samples (page 11)

Q12: Hydrogen

Q13: Helium

Q14: Calcium, since 650 nm has a triplet.

Q15: Sodium, as evidenced by the doublet at 580 nm.

Q16: Thallium

Q17: No, since there is no triplet at 580 nm.

Q18: No, since there is no triplet at 580 nm.

Q19: Thallium, since the line in sample B occurs at 530 nm and can only be credited to thallium from this database.

Obtaining wavelength values (page 14)

Q20:

1. The wavelength of light needed:

    Since $E = \dfrac{Lhc}{\lambda}$

    and $E = 338$ kJ mol^{-1}

    $$E = \frac{6.02 \times 10^{23}\text{mol}^{-1} \times 6.63 \times 10^{-34} \text{ J s} \times 3 \times 10^8 \text{ m s}^{-1}}{\lambda}$$

    $$\lambda = \frac{6.02 \times 10^{23}\text{mol}^{-1} \times 6.63 \times 10^{-34} \text{ J s} \times 3 \times 10^8 \text{m s}^{-1}}{338 \times 10^3 \text{ J mol}^{-1}}$$

    $\lambda = 0.3543 \times 10^{-6}$ m

    $\lambda = 354.3$ nm

    A major error is omitting multiplication by Avogadro's number.
    Remember that there are one mole of bonds.

2. These molecules can be unstable because this wavelength is within the ultraviolet region and sunlight, particularly in the upper atmosphere can provide this wavelength.

Energy from wavelength (page 15)

Q21: 74.8 kJ mol^{-1}

Q22: 92.1 kJ mol^{-1}

Q23: 85.5 kJ mol^{-1}

Q24: 63.0 kJ mol^{-1}

End of Topic 1 test (page 17)

Q25: a) γ - radiation

Q26: d) photons.

Q27: b) Colour moves towards red.

Q28: c) higher energy.

Q29: c) c

Q30: c) c and e) h, are both used to represent a constant

Q31: a) f

Q32: D

Q33: 408 nm

Q34: Green

Q35: 224 kJ mol^{-1}

Topic 2: Atomic orbitals, electronic configurations and the periodic table

Calculating ionisation energy (page 23)

Q1:
For one mole of electrons:
$$E = \frac{Lhc}{\lambda}$$
$$= \frac{6.02 \times 10^{23} \times 663 \times 10^{-34} Js \times 3 \times 10^8 ms^{-1}}{91.2 \times 10^{-9} m}$$
$$= 1.313 \times 10^6 J$$
$$= 1313 \text{ kJ mol}^{-1}$$

Relating quantum numbers (page 26)

Q2:

Value of n	Value of l	Value of m	Subshell name
1	0	0	1s
2	0	0	2s
	1	-1 0 +1	2p
3	0	0	3s
	1	-1 0 +1	3p
	2	-2 -1 0 +1 +2	3d

Spectroscopic notation (page 30)

Q3: c) $1s^2\ 2s^1$

Q4: d) Potassium

Q5: 4

Q6: 3

Q7: 1

Q8: d) Magnesium

Q9: b) $1s^2$

Orbital box notation (page 33)

Q10: Electron configuration

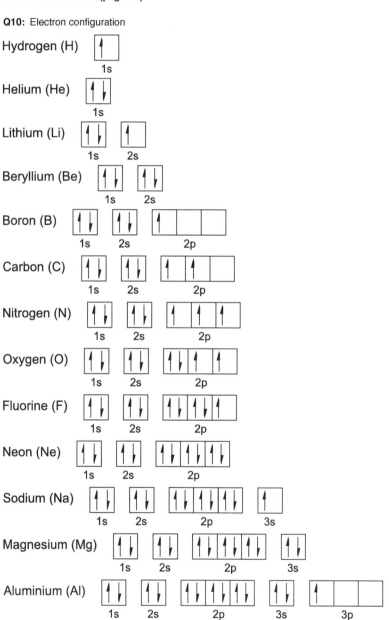

ANSWERS: UNIT 1 TOPIC 2

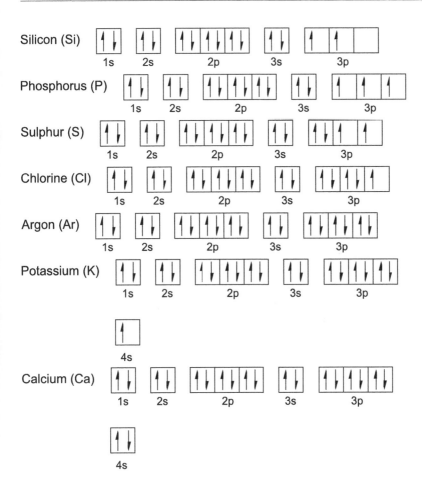

Periodic table blocks (page 34)

Q11:

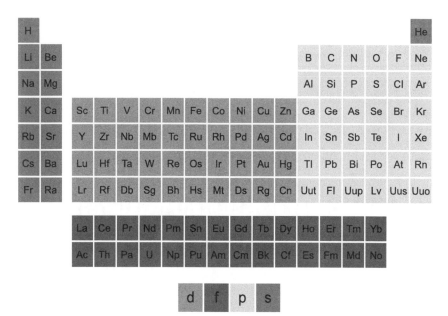

Q12: b) *p*

Q13: a) *s*

Q14: Transition

Q15: p

Q16: p

Q17: d

Q18: s

Ionisation energy evidence (page 36)

Q19: One

Q20: Easy

Q21: Reactive

Q22: One

First and second ionisation energies (page 38)

Q23: Sodium has **one electron** in its outer shell whereas magnesium has **two electrons** in its outer shell. The first ionisation energy of magnesium is **higher** than that of sodium since magnesium has 12 protons in its nucleus and therefore has a higher nuclear charge and a **stronger** attraction for the outer electrons. However, the second ionisation energy of **sodium** is higher than that of **magnesium** since the electrons being removed come from a **complete** p subshell which is **closer** to the nucleus.

Covalent bonding (page 39)

Q24: D

Q25: Bond length

Q26: Exothermic

Q27: The covalent radius for hydrogen is given as 37 pm and this is exactly half the bond length of the H-H bond since there are two hydrogens within it.

Q28: 136 pm since the covalent radius of chlorine is given as 99 pm and 99 + 37 = 136pm.

Resonance structures (page 41)

Q29: a) Cl_2

Q30: b) FCl

Q31: Methane and carbon dioxide have the following Lewis electron dot structures:

$$H \overset{\overset{H}{\underset{x}{\bullet}}}{\underset{\underset{H}{\overset{\bullet}{x}}}{\underset{x}{\bullet}C\overset{x}{\underset{\bullet}{\bullet}}}} H \qquad \overset{\overset{x\ \ x}{\bullet\bullet}}{\underset{\underset{\bullet\bullet}{\bullet}}{O}} \overset{\bullet}{\underset{\bullet}{\overset{x}{\underset{x}{C}}}} \overset{\overset{x\ \ x}{\bullet\bullet}}{\underset{\bullet\bullet}{O}}$$

Q32: Carbon monoxide (showing shared pairs as straight lines) is:

ANSWERS: UNIT 1 TOPIC 2

Summary of shapes of covalent molecules (page 47)

Q33:

Number of Electron Pairs	Arrangement	Angle(s) in degrees	Example
2	linear	180	$BeCl_2$
3	trigonal planar	120	BF_3
4	tetrahedral	109.5	CH_4
5	trigonal bipyramidal	90, 120, 180	PCl_5
6	octahedral	90	SF_6

Q34: a) PF_3

Q35: c) Trigonal bipyramid

Q36: d) BeF_2

Q37:

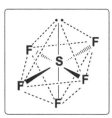

Q38: Octahedral

Q39: This shape is a square pyramid.

End of Topic 2 test (page 49)

Q40: a) the Pauli exclusion principle.

Q41: d) Chlorine

Q42: d) 4

Q43: c)

Q44: c) 3

Q45: b) would be needed to remove 1 mole of $2p$ electrons and 1 mole of $2s$ electrons.

Q46: a) $Mg^+(g) \rightarrow Mg^{2+}(g) + e^-$

Q47: 3 electrons

Q48: 1 electron

Q49: 3 quantum numbers

Q50: Degenerate

Q51: Argon

Q52: As the principal quantum number (n) increases, the energy levels become closer and closer together.

Q53: A phosphorus atom has three electrons in the $3p$ sub-shell, i.e. a half filled subshell which is relatively stable. A sulfur atom has four electrons in the $3p$ subshell, two unpaired and two paired. It is easier to remove one of the paired electrons from the sulfur atom than it is to remove an electron from the relatively stable phosphorus atom.

Q54: d) Calcium oxide

Q55: d) phosphorus donating both electrons of the bond to boron.

Q56: c) H_2S

Q57: b) Trigonal planar to tetrahedral

Q58: b) 107°

Q59: b) Angular

Q60: d) Trigonal pyramidal

Q61: d) Trigonal pyramidal

Q62: a) Tetrahedral

Q63: c) Three negative

Q64: a) A

Q65:

Dative covalent bond

Q66: b)

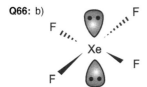

Q67: The arrangement of the bonding pairs is square planar.

Topic 3: Transition metals

Orbital box notation (page 58)

Q1:

Elements	Z	Electron configuration	Electron in box diagram
Scandium	21	$1s^2\ 2s^2\ 2p^6\ 3s^2\ 3p^6\ 4s^2\ 3d^1$	4s [↑↓] 3d [↑][ ][ ][ ][ ]
Titanium	22	$1s^2\ 2s^2\ 2p^6\ 3s^2\ 3p^6\ 4s^2\ 3d^2$	4s [↑↓] 3d [↑][↑][ ][ ][ ]
Vanadium	23	$1s^2\ 2s^2\ 2p^6\ 3s^2\ 3p^6\ 4s^2\ 3d^3$	4s [↑↓] 3d [↑][↑][↑][ ][ ]
Chromium	24	$1s^2\ 2s^2\ 2p^6\ 3s^2\ 3p^6\ 4s^1\ 3d^5$	4s [↑] 3d [↑][↑][↑][↑][↑]
Manganese	25	$1s^2\ 2s^2\ 2p^6\ 3s^2\ 3p^6\ 4s^2\ 3d^5$	4s [↑↓] 3d [↑][↑][↑][↑][↑]
Iron	26	$1s^2\ 2s^2\ 2p^6\ 3s^2\ 3p^6\ 4s^2\ 3d^6$	4s [↑↓] 3d [↑↓][↑][↑][↑][↑]
Cobalt	27	$1s^2\ 2s^2\ 2p^6\ 3s^2\ 3p^6\ 4s^2\ 3d^7$	4s [↑↓] 3d [↑↓][↑↓][↑][↑][↑]
Nickel	28	$1s^2\ 2s^2\ 2p^6\ 3s^2\ 3p^6\ 4s^2\ 3d^8$	4s [↑↓] 3d [↑↓][↑↓][↑↓][↑][↑]
Copper	29	$1s^2\ 2s^2\ 2p^6\ 3s^2\ 3p^6\ 4s^1\ 3d^{10}$	4s [↑] 3d [↑↓][↑↓][↑↓][↑↓][↑↓]
Zinc	30	$1s^2\ 2s^2\ 2p^6\ 3s^2\ 3p^6\ 4s^2\ 3d^{10}$	4s [↑↓] 3d [↑↓][↑↓][↑↓][↑↓][↑↓]

Electronic configuration (page 58)

Q2: Scandium only forms 3+ ions and zinc only forms 2+ ions. Neither of these result in an incomplete d subshell, therefore do not fit the definition of a transition metal.

Q3: Fe^{3+} ions would have a half-filled d subshell which is stable.

Calculating an oxidation state (page 59)

Q4: +5

Q5: +6

Q6: +4

Q7: +3

Q8: +6

Q9: +2

Q10: +6

Q11: +6

Q12: +3

Naming transition metal complexes (page 64)

Q13: Hexaaquacobalt (II) chloride

Q14: Sodium tetrafluoridochromate (III)

Q15: Potassium hexacyanidoferrate (III)

Q16: Potassium trioxalatoferrate (III)

Q17: 4

Q18: 6

Q19: 6

Q20: Octahedral

Q21: b) $Na_2[PtCl_4]$

Q22: d) $[Cu(CN)_2(OH_2)_2]$

Q23: a) $[CrCl(OH_2)_5]Cl_2$

Q24: c) $[CoCl_2(NH_3)_4]Cl$

Colour of transition metal compounds (page 67)

Q25: If violet is absorbed, **yellow-green** is transmitted.
If **purple** is absorbed, green is transmitted.
If orange is absorbed, **green-blue** is transmitted.
When all colours of light are present **white** light is produced.

Q26: Green-blue

Q27: Green-blue

Q28: Yellow-green

Explanation of colour in transition metal compounds (page 70)

Q29: 164.0

Q30: b) Blue-green

Q31: 210.1

Q32: c) Violet

ANSWERS: UNIT 1 TOPIC 3

Q33: 257.5

Q34: b) Yellow

Q35: c) $Cl^- < H_2O < NH_3$

End of Topic 3 test (page 77)

Q36: a) $1s^2\ 2s^2\ 2p^6\ 3s^2\ 3p^6\ 3d^3\ 4s^2$

Q37: d)

↑	↑	↑	↑	↑		↑

Q38: d) Cr^{3+}

Q39: b) +2

Q40: a) Green

Q41: b) a reduction with gain of one electron.

Q42: b) The concentration of the absorbing species can be calculated from the intensity of the absorption.

Q43: a) Cation and d) Octahedral

Q44: f) Monodentate

Q45: Hexaamminechromium(III) chloride

Q46: The oxidation state of cobalt in this complex is +3.

Q47: Ammonia (NH3) causes the stronger ligand field splitting.

Q48: The reaction speeds up when the cobalt(II) chloride is added.

Q49: c) Cobalt exhibits various oxidation states of differing stability.

Q50: $[Co(NH_3)_6]^{3+}$ ions are yellow (red and green mixed) which means that they must absorb blue light.

$[CoF_6]^{3-}$ ions are blue which means that they must absorb yellow light (red and green mixed).

Blue light is of higher energy than yellow light. So, ammonia ligands produce a greater splitting of the d orbitals than fluoride ions.

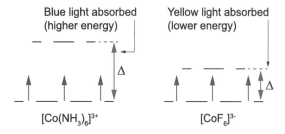

© HERIOT-WATT UNIVERSITY

Topic 4: Inorganic Chemistry test
Inorganic Chemistry test (page 82)

Q1: a) lower frequency.

Q2: b) the energy change when an electron moves to a lower energy level.

Q3: Movement of electrons from higher to lower energy levels.

Q4: 5.09×10^{14} Hz

Q5: a) $1s^2 2s^2 2p^6 3s^2 3p^6 3d^3 4s^2$

Q6: c) Hund's rule

Q7:

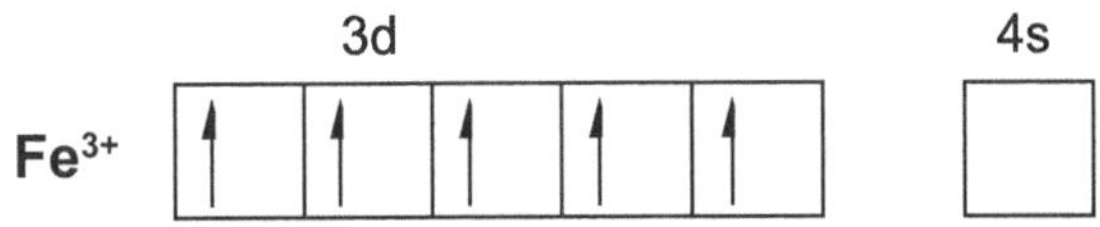

Fe^{3+} — 3d: ↑ ↑ ↑ ↑ ↑ ; 4s: (empty)

Q8: Stability of having all d orbitals half-filled.

Q9: a) Trigonal pyramidal

Q10: a) Tetrahedral

Q11: 2

Q12: 2

Q13: a) angular

Q14: 3

Q15: 1

Q16: d) trigonal pyramidal

Q17: 4

Q18: 0

Q19: f) tetrahedral

Q20: b) $[Ti(NH_3)_6]^{3+}$

Q21: Tetrachloridocuprate (II)

Q22: c) +3

Answers to questions and activities for Physical Chemistry

Topic 1: Chemical equilibrium

Hydrogen Iodide equilibrium (page 92)

Q1: At t = 15 there are 5, 5 and 2, respectively.
At t = 30 and t = 70 there is no change with four of each.

Q2: There are four of each at both times.

Q3: They are the same.

Equilibrium expressions 1 (page 95)

Q4: $K_c = \dfrac{[Fe^{2+}]^2[I_3^-]}{[Fe^{3+}]^2[I^-]^3}$

Q5: $K_c = \dfrac{[H^+]^2[HPO_4^{2-}]}{[H_3PO_4]}$

Equilibrium reactions (page 96)

Q6: c) Hydrogen iodide

Q7: a) Phosphorus(V) chloride

Q8: b) Mg

Q9: d) $2SO_2(g) + O_2(g) \rightleftharpoons 2SO_3(g)$ K_c at 636°C = 3343

Q10: The value increases from 21.1 to 3343 as the temperature drops.

Q11: a) 636°C

Equilibrium expressions 2 (page 97)

Q12: $Kp = \dfrac{(pNO)^2(pCl_2)}{(pNOCl)^2}$

Q13: $Kp = \dfrac{(pSO_3)^2}{(pSO_2)^2(pO_2)}$

Paper chromatography (page 101)

Q14: b) Black

Q15: e) Green

Q16: a) has the highest solvent/water partition coefficient.

Q17: c) Dark blue

Q18: They are probably the same dyes in both cases since the R_f values would be the same. If they were different materials, they would probably have moved different distances.

Calculating pH (page 106)

Q19: 2.30 pH

Q20: 5.10 pH

Q21: 12.80 pH

Q22: 10.46 pH

Q23: $[H_3O^+] = 5.00 \times 10^{-3}$ mol ℓ^{-1}
$[OH^-] = 2.00 \times 10^{-12}$ mol ℓ^{-1}

Q24: $[H_3O^+] = 2.51 \times 10^{-6}$ mol ℓ^{-1}
$[OH^-] = 3.98 \times 10^{-9}$ mol ℓ^{-1}

Q25: $[H_3O^+] = 3.98 \times 10^{-12}$ mol ℓ^{-1}
$[OH^-] = 2.51 \times 10^{-3}$ mol ℓ^{-1}

Q26: $[H_3O^+] = 1.26 \times 10^{-2}$ mol ℓ^{-1}
$[OH^-] = 7.94 \times 10^{-13}$ mol ℓ^{-1}

Strong/weak acids and bases (page 111)

Q27: 2.88

Parent acid and base (page 112)

Q28: The acid is nitric acid and the base is magnesium hydroxide.

Q29: The acid is hydrobromic acid and the base is potassium hydroxide.

Q30: The acid is ethanoic acid and the base is sodium hydroxide.

Q31: The acid is sulfurous acid (not sulfuric) and the base is calcium hydroxide.

Salts (page 112)

Q32: c) 9

Q33: Alkaline

Q34: d) The acid, hydrogen cyanide, is weak and the base is strong.

Q35: 9

Q36: If the pH is less than 7, then the acid is stronger than the base. So pyridine must be a weak base.

pH titration (page 116)

Q37: 7

Q38: At the equivalence point, the exact amount of alkali has been added to neutralise the acid; no more, no less.

Q39: 0.1

Q40: 3

Q41: d) The pH changes rapidly only around the equivalence point.

Titration curves (page 117)

Q42: a) Strong acid/strong alkali

Q43: c) 7

Q44: b) 5

Q45: d) 9

Q46: The pH at the equivalence point is the same as the pH of the salt formed.

combination	pH of salt
strong acid/strong alkali	7
strong acid/weak alkali	<7
weak acid/strong alkali	>7
weak acid/weak alkali	depends on relative strengths

Choosing indicators (page 118)

Q47: a) Suitable

Q48: a) Suitable

Q49: b) Unsuitable

Q50: b) Bromothymol blue

Q51: Bromothymol blue changes colour over the pH range 6.0-7.6 which contains the equivalence point. Phenolphthalein will also work well since the pH of the solution is changing rapidly over its pH range of around 8.0-10.0 For both these indicators adding a single drop of alkali at the end point should cause the colour change. Methyl orange is less suitable since a larger volume would be needed to cause its colour to change.

Q52: b) Unsuitable

ANSWERS: UNIT 2 TOPIC 1

Q53: a) Suitable

Q54: a) Suitable

Q55: d) Either methyl orange or bromothymol blue

Q56: For both methyl orange and bromothymol blue the pH of the solution is changing rapidly over the indicators' pH ranges. So there will be a sharp endpoint even although the equivalence point falls in neither range.

Q57: a) Suitable

Q58: b) Unsuitable

Q59: b) Unsuitable

Q60: a) Phenolphthalein

Q61: The pH of the equivalence point falls within the pH range over which phenolphthalein changes colour. So there will be a sharp endpoint. Both the other indicators will change colour gradually. For methyl orange, the colour change takes place long before the equivalence point.

Q62: b) Unsuitable

Q63: b) Unsuitable

Q64: b) Unsuitable

Q65: d) None of these

Q66: The pH change around the equivalence point is fairly gradual. In general, no indicator is suitable for the titration of a weak acid and a weak alkali. Such titrations have to be monitored using a pH meter.

Summary of buffer systems (page 127)

Q67:
Adding acid H_3O^+ + $\boxed{NH_3}$ $\rightleftharpoons$ $\boxed{NH_4^+}$

Adding alkali OH^- + $\boxed{NH_4^+}$ $\rightleftharpoons$ $\boxed{NH_3}$ + $\boxed{H_2O}$

Q68:
Adding acid H_3O^+ + $\boxed{CH_3COO^-}$ $\rightleftharpoons$ $\boxed{CH_3COOH}$

Adding alkali OH^- + $\boxed{CH_3COOH}$ $\rightleftharpoons$ $\boxed{CH_3COO^-}$ + $\boxed{H_2O}$

Buffer calculations (page 130)

Q69: 5.46

Q70: 1.78

Q71: 1.79×10^{-5}

© HERIOT-WATT UNIVERSITY

Q72: 4.6

Q73: 4.51

Q74: 1.55×10^{-5}

Extra questions (page 131)

Q75: c) CH_3COO^-

Q76: d) HSO_4^-

Q77: d) H_2O

Q78: a) Greater

Q79: b) Decreases.

Q80: 7.00

Q81: 3.51

Q82: 7.40

Q83: a) $K_c = H_3O^+] [OH^-] / [H_2O]$

Q84: c) $K_w = [H_3O^+] [OH^-]$

Q85: In **neutral water** the concentrations of the ionic species $[H_3O^+]$ is **equal to** $[OH^-]$ and the value of K_w is 1.0×10 **-14** at 25°C.

In acidic solutions the concentrations of the ionic species $[H_3O^+]$ is **greater than** $[OH^-]$ and the value of K_w is 1.0×10 **-14** at 25°C.

In alkaline solution the concentrations of the ionic species $[H_3O^+]$ is **less than** $[OH^-]$ and the value of K_w is 1.0×10 **-14** at 25°C.

Q86: 3.3

Q87: 1.6×10^{-10}

Q88: DBCA

Q89: CADB

Q90: Acidic

Q91: Less

Q92: Greater

Q93: b) $H_2PO_4^-$

Q94: Phosphoric acid, H_3PO_4

Q95: b) Basic

Q96: Ammonia

Q97: Ammonium ion

Q98: a) Ammonia, NH_3

Q99: b) Ammonium ion, NH_4^+

Q100: An **acidic** buffer solution contains a mixture of a **weak acid** and one of its **salts**. An example is a mixture of **ethanoic acid** and potassium ethanoate in water.
A **basic** buffer solution contains a mixture of a **weak base** and one of its **salts**. An example is a mixture of **ammonia** and ammonium chloride in water.

Q101: Relative formula mass of ethanoic acid.

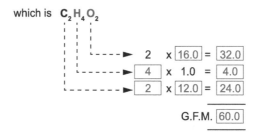

Q102: 6.0

Q103: 1.72×10^{-5}

Q104: 5.07

Q105: 4.47

Q106: 0.27

Q107: b) 100 cm^3 of 0.1 mol ℓ^{-1} HCOOH/0.2 mol ℓ^{-1} HCOO$^-$Na$^+$

Q108: c) Stay the same

End of Topic 1 test (page 142)

Q109: a) The reaction is endothermic.

Q110: b) $Cu(s) + Mg^{2+}(aq) \rightleftharpoons Cu^{2+}(aq) + Mg(s)$

Q111: d) Increase of temperature

Q112: b) 8×10^{-2} atm

Q113: c) $\dfrac{[NH_3]^2}{[H_2]^3[N_2]}$

Q114: 1.8 mol l^{-1}

Q115: 1.4 mol l^{-1}

Q116: K = 0.032

Q117: The forward reaction is exothermic so when the temperature is raised the reaction will go in reverse to absorb heat (le Chatelier's principle). The value of K at this increased temperature will, therefore, be reduced.

Q118: d) mass of solute involved.

Q119: d) equal.

Q120: a) 0.5

Q121: 0.015 mol l^{-1}

Q122: 0.0075 mol l^{-1}

Q123: K = 2

Q124: c) C

Q125: c) C

Q126: b) Hexane

Q127: b) $H_2O(l) + NH_3(aq) \rightarrow NH_4^+(aq) + OH^-(aq)$

Q128: d) NH_4^+ is the conjugate acid of NH_3.

Q129: c) 4.7

Q130: a) and d)

Q131: b) $K_w = [H_3O^+][OH^-]$

Q132: d) water will have a greater electrical conductivity at 25°C than at 18°C.

Q133: b) 0.50 mol l^{-1}

Q134: c) 10^{-10} and 10^{-11} mol l^{-1}

Q135: 3.4

Q136: 3.1

Q137: c) The overall colour of the solution depends on the ratio of [HIn] to [In⁻].

Q138: b) yellow in a solution of pH 3 and blue in a solution of pH 5.

Q139: c)
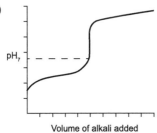

Q140: c) Phenolphthalein indicator, pH of colour change 8.0-9.8

Q141: a) The pH changes gradually around the equivalence point.

Q142: c) weak acid and a salt of that acid.

Q143: b) 50 cm³ NH₄Cl (aq) + 50 cm³ NH₃ (aq)

Q144: 3.47

Q145: 0.025 mol l⁻¹

Q146: c) Phenol red indicator, pH range 6.8 - 8.4

Q147: The salt sodium methanoate is formed in the reaction. Between E and F, some methanoic acid still remains unreacted and so there is a mixture of methanoic acid and sodium methanoate. A mixture of a weak acid and a salt of that acid is a buffer solution.

Q148: The methanoate ions present are able to remove added H_3O^+ ions. This creates methanoic acid.

Topic 2: Reaction feasibility

Calculating entropy in a chemical reaction (page 156)

Q1: $\Delta S° = \Sigma S°_{(products)} - \Sigma S°_{(reactants)}$
$= [(2 \times 43) + (2 \times 241) + 205] - (2 \times 142)$
$= 773 - 284$
$= 489$ J K^{-1} mol^{-1}

The entropy values for Ag and O_2 are found on page 17 of the CFE Higher and Advanced Higher data book.

The entropy values for $AgNO_3$, NO_2 and Ag are multiplied by two due to the ratio of moles within the chemical equation.

The second law of thermodynamics (page 157)

Q2: At 25°C (298 K), ΔS(total) is negative (-436 J K^{-1} mol^{-1})
At 1500°C (1773 K), ΔS(total) is positive (+60.6 J K^{-1} mol^{-1})

Q3: At 25°C (298 K), ΔS(total) = negative (approx -10 J K^{-1} mol^{-1})
At 5000°C (5273 K), ΔS(total) = negative (approx -3.7 J K^{-1} mol^{-1})
It is not thermodynamically feasible at either temperature.

Calculations involving free energy changes (page 160)

Q4: a) $2Mg(s) + CO_2(g) \rightarrow 2MgO(s) + C(s)$

$$\Delta G° = \Sigma G°_{PRODUCTS} - \Sigma G°_{REACTANTS}$$
$$= (2 \times -569 + 0) - (2 \times 0 + (-394))$$
$$= -744 \text{ kJ mol}^{-1}$$

The reaction where magnesium reduces carbon dioxide is feasible under standard conditions.

b) $2CuO(s) + C(s) \rightarrow 2Cu(s) + CO_2(g)$

$$\Delta G° = \Sigma G°_{PRODUCTS} - \Sigma G°_{REACTANTS}$$
$$= (2 \times 0 + (-394)) - ((2 \times -130) + 0)$$
$$= -134 \text{ kJ mol}^{-1}$$

The reaction where carbon reduces copper(II) oxide is feasible under standard conditions.

Q5: $\Delta H° = +117$ kJ mol^{-1}

$\Delta S° = +175$ J K^{-1} mol^{-1}
Since $\Delta G° = \Delta H° - T\Delta S°$
At 400 K $\Delta G°_{400} = +47$ kJ mol^{-1}
At 1000 K $\Delta G°_{1000} = -58$ kJ mol^{-1}

This reaction is feasible only at higher temperatures.

Q6:

$\Delta H^o = -92.8 \times 10^3$ J mol^{-1}
$\Delta S^o = -198.6$ J K^{-1} mol^{-1}
when $\Delta G^o = 0$

$$T = \frac{\Delta H^o}{\Delta S^o}$$

T = 467.3 K

Q7:

a) ΔG° = +0.16 kJ
b) Equilibrium position favours the reactants.

Q8: Boiling point = 333 K or 60°C

Interpreting Ellingham diagrams 1 (page 164)

Q9: Silver(I) oxide. At 1000 K, ΔG is +60 kJ mol^{-1} on the graph. Even with no other chemical involved, this reverses to breakdown silver(I) oxide with ΔG° = -60 kJ mol^{-1}

Q10: Above 2200 K approximately. This would allow ΔG to be negative for:

2ZnO → 2Zn + O$_2$

Q11: a)

(i) At 1000 K the target equation is:

2C(s) + 2ZnO(s) → 2Zn(s) + 2CO(g)
(ii) 2C(s) + O$_2$(g) → 2CO(g) ΔG° = -400 kJ mol^{-1}
      2Zn(s) + O$_2$(g) → 2ZnO(s) ΔG° = -500 kJ mol^{-1}

(iii) Reverse the zinc equation:
2ZnO(s) → 2Zn(s) + O$_2$(g) ΔG° = +500 kJ mol^{-1}

(iv) Adding to the carbon equation gives:
2C(s) + O$_2$(g) + 2ZnO(s) → 2CO(g) + 2Zn(s) + O$_2$(g)
The oxygen on each side cancels out giving:
2C(s) + 2ZnO(s) → 2CO(g) + 2Zn(s)
ΔG° = +100 kJ mol^{-1}
So at 1000 K the reaction is **not** feasible.

b)

(i) At 1500 K the target equation is the same:

$2C(s) + 2ZnO(s) \rightarrow 2Zn(s) + 2CO(g)$

(ii) $\quad 2C(s) + O_2(g) \rightarrow 2CO(g) \qquad \Delta G° = -500$ kJ mol^{-1}

$\quad 2Zn(s) + O_2(g) \rightarrow 2ZnO(s) \qquad \Delta G° = -300$ kJ mol^{-1}

(iii) Reverse the zinc equation:

$2ZnO(s) \rightarrow 2Zn(s) + O_2(g) \qquad \Delta G° = +300$ kJ mol^{-1}

(iv) Adding to the carbon equation gives:

$2C(s) + O_2(g) + 2ZnO(s) \rightarrow 2CO(g) + 2Zn(s) + O_2(g)$

The oxygen on each side cancels out giving:

$2C(s) + 2ZnO(s) \rightarrow 2CO(g) + 2Zn(s)$

$\Delta G° = -200$ kJ mol^{-1}

So at 1500 K the reaction is feasible.

Q12: Where the two lines cross, $\Delta G° = 0$. Above this temperature, the reaction is feasible. Approximately 1200K.
Remember: The lower of the two lines operates as written and the upper line will be reversed.

Q13: As the zinc melts, the disorder (entropy) increases. Since the gradient is given by $-\Delta S$ (from the straight line $\Delta G = -T\Delta S + \Delta H$), the slope of the line changes.

Q14: Zinc vaporises at 1180 K with an increase in entropy and a subsequent change in the gradient of the line on the Ellingham diagram.

Interpreting Ellingham diagrams 2 (page 166)

Q15:

a) $2FeO(s) + 2C(s) \rightarrow 2Fe(s) + 2CO(g)$
b) $\Delta G° = -155$ kJ mol^{-1}
c) Above 1010 K
d) Below 980 K
e) It is a gas and can mix better with the solid iron(II) oxide.

Q16:

a) above about 2100 K
b) The high cost of maintaining temperature. The fact that magnesium is a gas at this temperature.
c) $\Delta G° = +160$ kJ mol^{-1} (there would be some leeway in this figure).
d) $\Delta G° = +68$ kJ mol^{-1} (dependent on your answer to part (c)).
e) Keeps the equilibrium following reaction from going in the reverse direction.

$$2MgO + Si \rightleftharpoons SiO_2 + 2Mg$$

ANSWERS: UNIT 2 TOPIC 2

End of Topic 2 test (page 169)

Q17: d) $N_2(g) + 3H_2(g) \rightarrow 2NH_3(g)$

Q18: c) -67.5 kJ mol^{-1} of ZnO

Q19: c) -242

Q20: c) large and negative.

Q21: c) 461 K

Q22: b) calcium.

Q23: The difference in the values of ΔG and ΔH is determined by the entropy change of the system. The second reaction involves a reactant, gaseous oxygen, with a high entropy forming a solid, aluminium oxide, with a low entropy.

Q24: A reaction with a negative value for $\Delta G°$ will occur spontaneously, this has a negative $\Delta G°$ value.

A positive $\Delta H°$ indicates an endothermic reaction that will take heat from the surroundings which will drop in temperature.

Q25: 10.1 J K^{-1}mol^{-1}

Q26: +34.3

Topic 3: Kinetics

Orders and rate constants (page 175)

Q1: Rate = k [Br$^-$][BrO$_3^-$][H$^+$]^2

Q2: 4

Q3: 2

Q4: 0

Q5: 1

Q6: 1

Q7: Rate = k [N$_2$O$_5$]

Q8: s^{-1}

Q9: 0.00044 (normal decimal form) **or** 4.4 × 10^{-4} (standard form)

Q10: 0.0000308 (normal decimal form) **or** 3.08 × 10^{-5} (standard form)

Q11: Rate = k [H$_2$O$_2$][I$^-$] **or** Rate = k [H$_2$O$_2$][I$^-$][H$^+$]^0

Q12: mol^{-1} ℓ s^{-1}

Q13: 0.023

Reaction mechanisms (page 178)

Q14: 3001.5

Q15: 30001.5 seconds.

The filler takes 30 s to fill one bottle, so the thousandth bottle will be filled after 1000 × 30 s and a further 1.5 s will be needed to cap and label it.

Questions on reaction mechanisms (page 179)

Q16: b) H$_2$S + Cl$_2$ → S + 2HCl

Q17: a) 4HBr + O$_2$ → 2H$_2$O + 2Br$_2$

Q18: b) False

Q19: b) False

Q20: a) True

Q21: b) False

ANSWERS: UNIT 2 TOPIC 3

Q22: b) False

Q23: a) True

Q24: $2H_2O_2 \rightarrow 2H_2O + O_2$

Q25: catalyst

Q26: intermediate

Q27: b) Rate = k $[H_2O_2][Br^-]$

End of Topic 3 test (page 182)

Q28: c) mol l^{-1} s^{-1}

Q29: a) Rate = k [X] [Y]

Q30: a) l mol^{-1} s^{-1}

Q31: d) the rate expression cannot be predicted.

Q32: d) 8

Q33: b) The rate of reaction decreases as the reaction proceeds.

Q34: c) X + Y $\rightarrow$ **intermediate**

Q35: d) k$[N_2O_5]$

Q36: c) P + Q $\rightarrow$ R + S slow
R + Q $\rightarrow$ T fast

Q37: a)

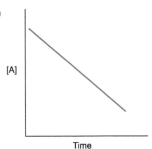

Q38: Rate = k $[C_2O_4^{2-}]2$ $[HgCl_2]$
One mark each for order with respect to each reactant.

Q39: 0.0154 mol^{-2} l^2 min^{-1}
1 mark for numerical answer, 1 mark for units

Q40: d) l^2 mol^{-2} min^{-1}

Q41: 1.54 × 10^{-5} mol l^{-1} min^{-1}

Topic 4: Physical chemistry test

Physical chemistry test (page 188)

Q1: b) decrease the concentration of SO_2 and O_2.

Q2: b) Hydrochloric acid and sodium chloride

Q3: Added hydroxide ions react with the hydrogen ions and ethanoic acid molecules further dissociate to replace these hydrogen ions.

$H^+ + OH^- \rightleftharpoons H_2O$ (l)

$CH_3COOH \rightleftharpoons H^+ + CH_3COO^-$

Q4: 4.86

Q5: $\Delta S° = 891$ J K^{-1} mol^{-1}

Q6: T= 572 K

Q7: d) 0 K

Q8: a) negative.

Q9: c) The reaction is second order overall.

Q10: Rate = $k[P][Q]^2$

Q11: P + 2Q → Intermediate

Q12: k = 0.15

Q13: d) l^2 mol^{-2} min^{-1}

Q14: b) 4

Answers to questions and activities for Researching Chemistry

Topic 3: Stoichiometric calculations

Excess calculations (page 200)

Q1: a) Zinc

Q2: a) Sulfuric acid

Nickel(II) ions and dimethylglyoxime (page 201)

Q3: 4

End of Topic 3 test (page 203)

Q4: d) phenolphthalein

Q5: Average titre in litres = ((first titration + second titration) / 2) / 1000 = ((8.7 + 8.6) / 2) / 1000 = 0.00865
Number of moles of sulfuric acid in the average titre = 0.00865 × 0.050 (concentration) = **4.325 × 10^{-4} moles**

Q6: Moles of sodium hydroxide in 25 cm^3 sample = 4.325 × 10^{-4} × 2 = 8.65 × 10^{-4} (reacts in a 2:1 ratio with the sulfuric acid)
Moles of sodium hydroxide in 250 cm^3 sample = 8.65 × 10^{-4} × 10 = **8.65 × 10^{-3} moles**

Q7: Starting moles of sodium hydroxide = 0.0025 × 1 = 0.025
Moles reacting = 0.025 - 8.65 × 10^{-3} = **0.01635 or 1.635 × 10^{-2} moles**

Q8: Moles of acetylsalicylic acid = 0.01635 / 2 = 0.008175 (molar ratio with sodium hydroxide from equation 1:2)
Mass of acetylsalicylic acid = 0.008175 × 180 = 1.4715 g (5 tablets)
Mass in one tablet = 1.4715 / 5 = **0.2943 g**

Q9: Calculate the mass of each element in the compounds formed from combustion of the compound as follows:

- Carbon = 3.52/44 × 12 = 0.96 g
- Hydrogen = 2.16/18 × 2 = 0.24 g
- Sulfur = 2.56/64.1 × 32.1 = 1.28 g

Molar ratio:

C = 0.96 /12 = 0.08 = **2**   H = 0.24 / 1 = 0.24 = **6**   S = 1.28 / 32.1 = 0.04 = **1**

Q10: 5.64 g of ethyl benzoate produces (5.64 × 122) / 150 = 4.59 g which is the theoretical yield.
Mass of benzoic acid produced = 73.2% of theoretical yield = 73.2% of 4.59 g = **3.36 g**

© HERIOT-WATT UNIVERSITY

Topic 4: Gravimetric analysis

Gravimetric analysis (page 208)

Q1: The compound is heated repeatedly until the mass does not change.

Q2: Prevent the compound from absorbing moisture (keep it dry).

Q3: Insoluble product is formed (precipitate formed).

Q4: Number of moles of nickel dimethylglyoxime (number of moles of nickel) = 0.942 / 288.7
Mass of nickel = n × GFM = (0.942 / 288.7) × 58.7 = 0.192 g
Percentage of nickel = 0·192 / 0·968 × 100 = ***19·8%***

Topic 5: Volumetric analysis

End of Topic 5 test (page 220)

Q1: n for benzoic acid = $16.55 \times 10^{-3} \times 0.0563 = 9.32 \times 10^{-4}$

20 cm^3 of NaOH was used. The number of moles of benzoic acid is equal to the number of moles of sodium hydroxide (a 1:1 ratio).

c for NaOH = $(9.32 \times 10^{-4}) / 0.020 = $ ***0.0466 mol l^{-1}***

Q2: Average titre volume = $(15.9 + 15.8) / 2 = 15.85$ cm^3

Number of moles of sulfuric acid used = $0.001585 \times 0.02 = 3.17 \times 10^{-4}$ mol

Number of moles of NaOH left in the 25.0 cm^3 hydrolysed solution = $2 \times 3.17 \times 10^{-4} = 6.34 \times 10^{-4}$ mol

Number of moles of NaOH left in 250.0 cm^3 of the hydrolysed solution = $10 \times 6.34 \times 10^{-4} = 6.34 \times 10^{-3}$ mol

Number of moles of NaOH added to aspirin initially = $0.025 \times 1 = 2.5 \times 10^{-2}$ mol

Number of moles NaOH reacted with aspirin = $2.5 \times 10^{-2} - 6.34 \times 10^{-3} = 1.866 \times 10^{-2}$ mol

2 moles of NaOH reacts with 1 mole aspirin, therefore number of moles of aspirin in 4 tablets = $1.866 \times 10^{-2} / 2 = 9.33 \times 10^{-3}$ mol

Number of moles of aspirin in 1 tablet = $9.33 \times 10^{-3} / 4 = 2.333 \times 10^{-3}$ mol

Mass of aspirin in each tablet = n × gfm = $2.333 \times 10^{-3} \times 180 = $ ***0.420 g***

Topic 6: Practical skills and techniques

End of Topic 6 test (page 235)

Q1: Make up several solutions accurately of various concentrations of permanganate and measure the absorbance of each one. *(1 mark)*

Draw a calibration curve of concentration versus absorbance. *(1 mark)*

The absorbance of the unknown permanganate concentration is measured and using the calibration curve the concentration of the solution is determined. *(1 mark)*

Q2: The exact mass must be in the region of 0.5 g and the exact mass must be known.

Q3: Pour mixture into standard flask, rinse beaker with water and add rinsings to the flask. *(1 mark)*

Make up to the mark with distilled water, stopper and invert flask. *(1 mark)*

Q4: Results that are ± 0.1 cm^3 in range from each other.

Q5: Low gram formula mass (GFM) or unstable in air or absorbs moisture or not a primary standard.

Q6: Solvent extraction

Q7:

1. Drain layers into separate beakers. *(1 mark)*
2. Return lower/aqueous layer to the separating funnel add more diethyl ether and repeat. *(1 mark)*
3. Evaporate/distill all diethyl ether layers. *(1 mark)*

Q8:

1. Immiscible in water. *(1 mark)*
2. Benzocaine is soluble within it. *(1 mark)*

Q9: Recrystallisation or chromatography

Q10: Pure benzocaine

Q11: It has a small impurity so is not 100% pure.

Topic 7: Researching chemistry test

Researching chemistry test (page 240)

Q1: Mass of water driven off = mass of hydrated barium chloride - mass of anhydrous barium chloride = 2.58 - 2.22 = 0.36 g

Number of moles of H_2O = 0.36 / 18 = 0.020 mol
Number of moles of $BaCl_2$ = 2.22 / 208.3 = 0.011 mol

Ratio of moles $BaCl_2:H_2O$ = 0.011 / 0.02 = 1:1.8 which is 1:2 to the nearest whole number.
Therefore ***n=2*** and the formula is $BaCl_2.2H_2O$

Q2: Number of moles of $CaSO_4$ = 3.89 / 136.2 = 0.0286

Number of moles of H_2O = (4.94 - 3.89) / 18 = 1.05 / 18 = 0.0583

Ratio of moles = 0.0286:0.0583 = 1:2 therefore ***x=2***.

Q3: Number of moles of sulfuric acid = v × c = 0.0178 (litres) × 0.22 = 0.003916 moles

Balanced stoichiometric equation shows that 2 moles of NaOH reacts with 1 mole of H_2SO_4

2NaOH +	H_2SO_4	→	$Na_2SO_4 + 2H_2O$
2 mol	1 mol		
2 × 0.003916	0.003916		

In the 25.0 cm³ sample of diluted drain cleaner there was 2 × 0.003916 = 0.007832 moles of NaOH.
In the 250.0 cm³ standard flask there would have been 0.007832 × 10 = 0.07832 moles of NaOH which is the number of moles in the 10.0 cm³ undiluted drain cleaner.

In 1 litre there would be 0.07832 × 100 = 7.832 moles of NaOH

Mass of NaOH = n × gfm = 7.832 × 40.0 = ***313 g in 1 litre***.

Q4: Acid hydrolysis

Q5: To purify the sulfanilamide

Q6:

1 mole of 4-acetamidobenzenesulfonamide	→	1 mole of sulfanilamide
214.1 g	→	172.1 g
4.282 g		(4.282 × 172.1)/214.1 = 3.442 g

Percentage yield = (actual yield / theoretical yield) × 100 = (2.237 / 3.442) × 100 = 0.65 × 100 = ***65%***

Q7: Sample is mixed with pure sulfanilamide and if pure the melting point of the mixture will be the same as the pure sample.

Q8: Any one of:
- thin-layer chromatography (TLC)
- nuclear magnetic resonance (NMR) spectroscopy
- infrared (IR) spectroscopy

Q9: Colourless to pink/purple

Q10: Number of moles of acidified potassium dichromate n = c × v = 0.02 × 0.0165 = 0.00033
Oxalate reacts in a 5:2 ratio so the number of moles of oxalate = 0.00033 × 2.5 = **0.000825** or **8.25 × 10^{-4}**

Q11: Mass in 1 litre = number of moles of oxalate × original conc in 20 cm^3 × gfm = 0.000825 × 50 × 88 = **3.63 g**

Q12: Mass of potassium = total mass - mass of oxalate - mass of hydrogen = 4.49 - 3.63 - 0.06 = **0.8 g**

Q13: K = 0.8 / 39 = 0.02
H = 0.06 / 1 = 0.06
C_2O_4 = 3.63 / 88 = 0.04
Ratio of x, y, z = 0.02:0.06:0.04 = 1:3:2
x=1, y=3 and z=2

Answers to questions and activities for Organic Chemistry and Instrumental Analysis

Topic 1: Molecular orbitals

Bonding in hydrocarbons (page 251)

Q1: c) sp

Q2: c) There are two π bonds and three σ bonds.

Q3: Linear

End of Topic 1 test (page 253)

Q4: d) 8 sigma bonds, 1 pi bond

Q5: The mixing of one s and three p orbitals.

Q6: d) 8

Q7: a) sp^2 hybridised carbon atoms but no sp^3 hybridised carbon atoms.

Q8: Electronegativity difference between Ti and Cl is 1.5 (value of 3.0 for chlorine and 1.5 for titanium), therefore the bonding is probably polar covalent and the bonding orbitals will lie around the chlorine atom due to it having the higher electronegativity.

Topic 2: Synthesis

The reaction of chlorine with benzene (page 287)

Q1: c) Electrophilic substitution

Q2: b) Chlorobenzene

Q3: d) Heterolytic

Q4: Carbocation

Di- and trinitrobenzene structures (page 288)

Q5: d) Electrophile

Q6: b) Carbocation

Q7: The intermediate is stabilised by delocalisation of the positive charge around the benzene ring.

Q8: c) 1,3-dinitrobenzene

Benzene (page 290)

Q9: Electrophile

Q10: Substitution

Methylbenzene (page 291)

Q11: Catalyst

Q12: c) Heterolytic

Q13: Chloroethane

Extra practice questions (page 292)

Q14:

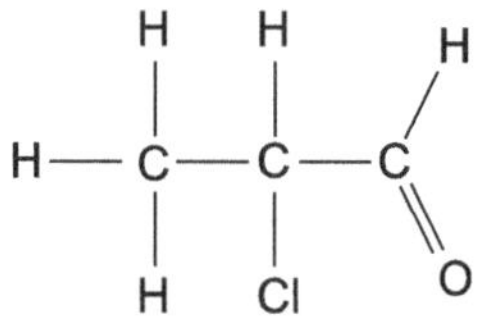

$$\text{H}-\underset{\underset{\text{H}}{|}}{\overset{\overset{\text{H}}{|}}{\text{C}}}-\underset{\underset{\text{Cl}}{|}}{\overset{\overset{\text{H}}{|}}{\text{C}}}-\text{C}\underset{\text{O}}{\overset{\text{H}}{\diagup\hspace{-6pt}\diagdown}}$$

Q15: Lithium aluminium hybride/LiAlH$_4$
Sodium borohybride/sodium Tetrahydroborate/NaBH$_4$

ANSWERS: UNIT 4 TOPIC 2

Q16: Electrophilic substitution or alkylation

Q17: Light/UV radiation

Q18:

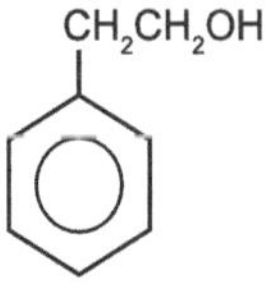

$C_6H_5-CH_2CH_2OH$ (phenyl group with CH_2CH_2OH substituent)

Q19: (Base-induced) elimination

Q20:

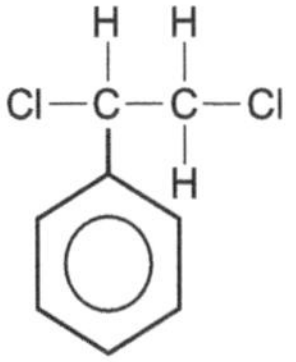

$C_6H_5-CHCl-CH_2Cl$ or benzene ring with $CHClCH_2Cl$ substituent

$$Cl-\underset{H}{\underset{|}{C}}-\underset{H}{\underset{|}{C}}-Cl$$ (with phenyl attached) or phenyl-$CHClCH_2Cl$

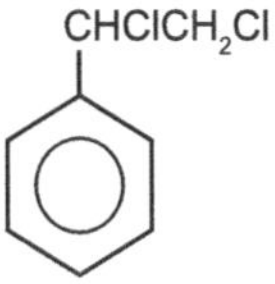

Q21: 2-hydroxypropanoic acid

Q22: Carbon atom 2 because 4 different groups are attached.

Q23: KCN or NaCN or HCN

Q24: Hydrolysis, acid hydrolysis

Q25: But-2-ene has two different groups attached to each of the carbon atoms of the double bond.

Q26:

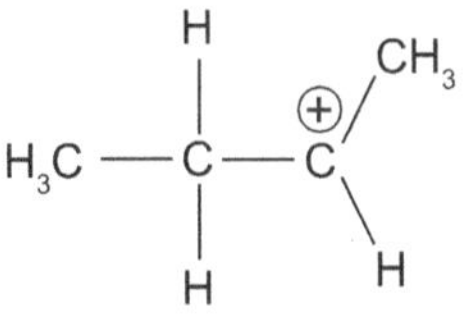

$$H_3C-\underset{H}{\underset{|}{C}}-\overset{+}{C}\diagup^{CH_3}_{\diagdown H}$$

Q27: Potassium (or sodium) hydroxide solution

Q28: Aluminium chloride

© HERIOT-WATT UNIVERSITY

Q29:

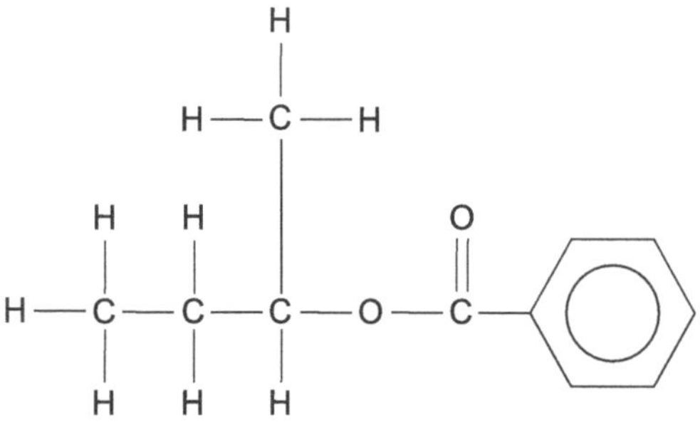

```
              H
              |
          H—C—H
   H   H   |       O
   |   |   |       ||
H—C—C—C—O—C—⬡
   |   |   |
   H   H   H
```

End of Topic 2 test (page 296)

Q30: a) ethene acting as an nucleophile and Br⁻ acting as an nucleophile.

Q31: Propane

Q32: 1,2-dibromopropane

Q33: b) Secondary

Q34: 1-ethoxy-2-methylpropane

Q35: Secondary

Q36: b) 2-methylbut-1-ene

Q37: An acid chloride

Q38: The reaction is too slow.

Q39: c) 3

Q40: c) The ethanoate ion is more stable than the ethoxide ion due to electron delocalisation.

Q41: a) $MgCO_3$

Q42: They are both salts

Q43: An amide

Q44: Propan-1-ol

Q45: Propyl propanoate

Q46: d) resists addition reactions.

Q47: b)

$$\text{C}_6\text{H}_5-\text{CH}(\text{CH}_3)_2$$

(benzene ring with -CH(CH₃)₂ substituent)

Q48: 1-bromo-3-methylbenzene

Q49: 3-chloro-4-methylphenol

Q50: 1,2-dimethylbenzene, 1,4-dimethylbenzene, ethylbenzene are all possible aromatic isomers. There are some elaborate isomers which are not aromatic and you may like to while away a few minutes finding some!

Q51: Alkylation and sulfonation

Q52: d) Nitration

Q53: b) 2

Q54: a) Primary

Q55: c) 2-aminobutane

Q56: Ethylamine and nitric acid (hydrogen nitrate) in the gas state give ethylammonium nitrate as small white crystals.

Q57: Ethylamine behaves as a proton acceptor (a base) and is neutralised by the acidic nitric acid. The reaction is a neutralisation.

Q58: The hydrogen of the nitric acid is attracted to the ethylamine which is basic. The hydrogen is therefore acting as an electrophile.

Q59: The lone pair on the ethylamine is attracted to the hydrogen of the nitric acid and the hydrogen of the nitric acid is therefore acting as an electrophile, the nitrogen a nucleophile.

Q60: The longer the alkyl chain in an amine, the lower the solubility in water. There is only one primary amine with a shorter alkyl chain than ethylamine, and that is methylamine.

Topic 3: Stereochemistry

Optical isomers of alanine (page 312)

Q1: a) Structure 1

Q2: b) non-superimposable.

Q3: b) asymmetric.

Q4: b) chiral.

End of Topic 3 test (page 316)

Q5: d) C_2F_4

Q6: d)

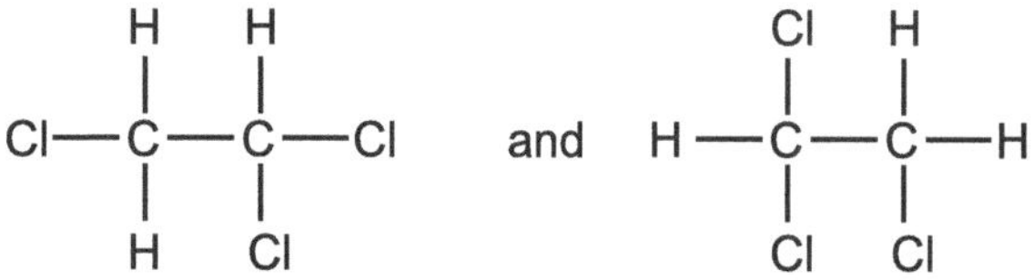

```
    H   H                    Cl  H
    |   |                    |   |
Cl—C — C—Cl     and     H—C — C—H
    |   |                    |   |
    H   Cl                   Cl  Cl
```

Q7: c) Butan-2-ol

Q8: b) 1-chloroethanol

Q9: C

Q10: A and C

Q11: A

Q12: D

Q13: D

Q14: A

Q15: c) sample tube of chiral molecules.

Q16: Y

Q17: Y

Q18: b) Two

Q19: Optical isomers are also called enantiomers.

Q20: b) Geometric isomers

Q21: a) Structural isomers

Q22: c) A and B

Q23: c)

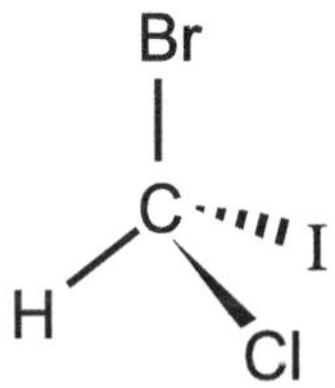

$$\text{H}-\overset{\overset{\displaystyle \text{Br}}{|}}{\text{C}}\cdots\text{I}$$
$$\text{Cl}$$

Q24: A and D

Topic 4: Experimental determination of structure

Calculate the empirical formula (page 325)

Q1: C_2H_6O

Questions about fragmentation (page 329)

Q2:

butan-1-ol $CH_3CH_2CH_2CH_2OH$

butan-2-ol $CH_3CH_2\underset{|}{\overset{\overset{\displaystyle CH_3}{|}}{C}}H\,OH$

2-methylpropan-2-ol $CH_3\underset{\underset{\displaystyle CH_3}{|}}{\overset{\overset{\displaystyle CH_3}{|}}{C}}OH$

2-methylpropan-1-ol $CH_3\underset{\underset{\displaystyle CH_3}{|}}{C}H\,CH_2\,OH$

methoxopropane
$$H-\underset{\underset{\displaystyle H}{|}}{\overset{\overset{\displaystyle H}{|}}{C}}-O-\underset{\underset{\displaystyle H}{|}}{\overset{\overset{\displaystyle H}{|}}{C}}-\underset{\underset{\displaystyle H}{|}}{\overset{\overset{\displaystyle H}{|}}{C}}-\underset{\underset{\displaystyle H}{|}}{\overset{\overset{\displaystyle H}{|}}{C}}-H$$

ethoxyethane
$$H-\underset{\underset{\displaystyle H}{|}}{\overset{\overset{\displaystyle H}{|}}{C}}-\underset{\underset{\displaystyle H}{|}}{\overset{\overset{\displaystyle H}{|}}{C}}-O-\underset{\underset{\displaystyle H}{|}}{\overset{\overset{\displaystyle H}{|}}{C}}-\underset{\underset{\displaystyle H}{|}}{\overset{\overset{\displaystyle H}{|}}{C}}-H$$

Q3: 74

Q4: CH_3

Q5: 2-methylpropan-2-ol has three methyl groups attatched to the carbon with the OH group attatched, and is likely most easily to lose one of them. Butan-2-ol also has a single CH_3 attatched to the carbon with the OH on, so would be expected to have a reduced m/z 59 ion. (Possibly spectrum Y)

Q6: Loss of the CH_3CH_2 group from the carbon with the OH, in a similar manner to loss of the CH_3 shown previously, will leave a fragment of this m/z value, with structure $[HC(CH_3)OH]^+$. This suggests further that spectrum Y is butan-2-ol.

Q7: $[CH_3CH_2CH_2]^+$ with m/z 43 and $[CH_2OH]^+$ with m/z 31.

Q8: m/z 31 is the base peak in spectrum Z, and m/z 43 is also a major peak. In other spectra they are less prominent. Since these two fragments are easily produced from the structure of butan-1-ol, spectrum Z applies to this.

Q9: Mass spectrum X is for 2-methylpropan-2-ol (tertiary alcohol). The peak at m/z 59 is due to the $[C(CH_3)_2OH]^+$ fragment. A $[CH_3]^+$ fragment adjacent to the C-OH has been lost from the molecule to give this fragment.

Mass spectrum Y is for butan-2-ol (secondary alcohol). The peak at m/z 45 is due to the $[CH_3CHOH]^+$ fragment and the peak at m/z 59 is due to the $[C_2H_5CHOH]^+$ fragment.

Mass spectrum Z is for butan-1-ol. A $[CH_3CH_2CH_2]^+$ fragment has been lost to give the peak at m/z 31 which is due to the $[CH_2OH]^+$ fragment.

Questions on bond vibration (page 332)

Q10: a) HCl

Q11: c) HI

Q12: a) HCl

Functional group identification (page 335)

Q13: b) Y

Q14: b) Y

Q15: Aldehyde

Q16: c) Benzonitrile (C_6H_5CN)

Determining structure using NMR (page 340)

Q17: 2

Q18: 5:3

Q19: d) Ar**H**

Q20: c) ArCH_3

Q21: Toluene, $C_6H_5CH_3$, with 5 aromatic Hs and 3 methyl Hs.

Q22: c) 3

Q23: b) 4:5:6

Q24: For the 6 CH_3, δ range 0.8 - 1.3 (actually found at 1.2);
for the 4 CH_2 - N, δ range 2.5 - 3.0 (actually found at 3.5);
for the 5 aromatic H, δ range 6.5 - 8.3 (actually found at 7.2).

Q25:

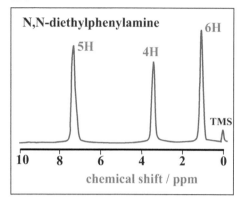

NMR - Practice (page 344)

Q26: The triplet at 1.058 is due to a CH_3 group adjacent to a CH_2 group. The singlet at 2.139 is due to a CH_3 group which is not adjacent to any other atoms with hydrogen atoms attached. The quartet at 2.449 is due to a CH_2CO group adjacent to a CH_3 group. This is also consistent with the ratio of the peak areas of 3:2:3. Therefore the molecule must contain two different CH_3 groups and a CH_2CO. This leads to the following structure:

$$CH_3CH_2\underset{\underset{O}{\|}}{C}CH_3$$

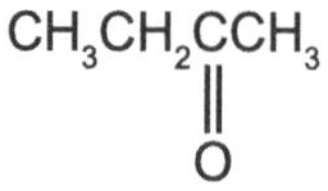

Q27: The triplet at 1.148 is due to a CH_3 group adjacent to a CH_2 group. The quartet at 2.324 is due to a CH_2CO group adjacent to a CH_3 group. The singlet at 3.674 is due to a CH_3-O group. This is also consistent with the ratio of the peak areas of 3:2:3. Therefore the molecule must contain a CH_3 group, a CH_2CO group and a CH_3-O group. This leads to the following structure:

$$CH_3CH_2C\begin{matrix}\diagup\!\!\!\!O\\ \diagdown OCH_3\end{matrix}$$

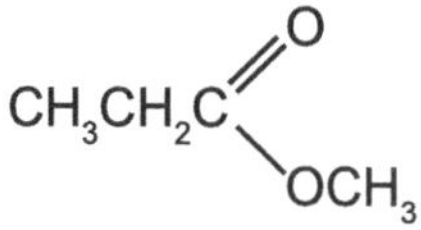

Q28: All three peaks here are singlets which means that none of the groups are adjacent to any atoms with hydrogen atoms attached. The ratio of the peak areas is consistent with two different CH_3 groups and a CH_2 group. The peak at 2.148 is due to a CH_3CO group. The peak at 4.029 is due to a CH_2-O group and the peak at 3.421 is due to a CH_3-O group. This leads to the following structure...

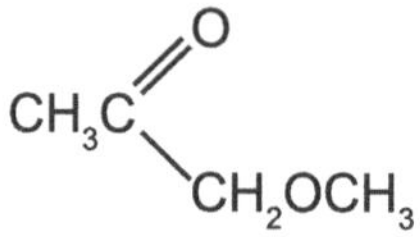

$$CH_3C\begin{matrix}\diagup\!\!\!\!O\\ \diagdown CH_2OCH_3\end{matrix}$$

ANSWERS: UNIT 4 TOPIC 4

End of Topic 4 test (page 351)

Q29: b) $[C_2H_4]^+$

Q30: $CH_3COOCH_2CH_2CH_2CH_3$

Q31: The molecular or parent ion, $[C_6H_{12}O_2]^+$

Q32: a) unstable and easily fragments to form daughter ions.

Q33: $[C_3H_7]^+$ or $[CH_3CO]^+$

Q34: $[CH_2OCOCH_3]^+$ or $[CH_3COOCH_2]^+$ or $[C_4H_9O]^+$; all three have m/z = 73.

Q35: 27

Q36: a) $[C_4H_9]^+$ at m/z 57

Q37: b) $[CHO]^+$ and e) $[C_2H_5]^+$

Q38: c) C_6H_6 and d) C_2H_2

Q39: a) $[CH_3]^+$

Q40: a) $[CH_3]^+$ and f) $[CH_2OH]^+$

Q41: c) $Si(CH_3)_4$

Q42: c)

$$CH_3CH_2-C\!\!\begin{array}{c}\diagup O \\ \diagdown H\end{array}$$

Q43: a) 1

Q44: There are three different hydrogen environments, therefore three peaks.

Q45: The peak areas are in the ratio 3:2:3.

Q46: Both are alcohols.

Q47: C is an aldehyde.

Q48: D is a ketone.

Q49: E is a carboxylic acid, with a broad band in the 3500 to 2500 cm^{-1} range and one in the 1725 to 1700 cm^{-1} range.

Q50: a) 1,2-dichloroethane, CH_2ClCH_2Cl and c) Ethanenitrile, CH_3CN

Q51: c) Ethanenitrile, CH_3CN

Q52: The NMR chemical shift for RCH_2CN is 2.5 - 2.0 ppm.

© HERIOT-WATT UNIVERSITY

Q53: This is an open-ended question and as such it does not have only one correct answer. A variety of answers are acceptable.

In your answer to this question, you may wish to discuss the following - Chromophores, conjugation, colour related to wavelength of light, $E = Lhc/\lambda$, absorption of light, complementary colour, HOMO to LUMO transitions.

Open-ended questions are always marked out of a maximum of three marks according to the following criteria:

- 3 marks: The maximum available mark would be awarded to a student who has demonstrated, at an appropriate level, a good understanding, of the chemistry involved. The student shows a good comprehension of the chemistry of the situation and has provided a logically correct answer to the question posed. This type of response might include a statement of the principles involved, a relationship or an equation, and the application of these to respond to the problem. This does not mean the answer has to be what might be termed an 'excellent' answer or a 'complete' one.

- 2 marks: The student has demonstrated, at an appropriate level, a reasonable understanding of the chemistry involved. The student makes some statement(s) which is/are relevant to the situation, showing that the problem is understood.

- 1 mark: The student has demonstrated, at an appropriate level, a limited understanding of the chemistry involved. The student has made some statement(s) which is/are relevant to the situation, showing that at least a little of the chemistry within the problem is understood.

- Zero marks should be awarded if: The student has demonstrated no understanding of the chemistry involved at an appropriate level. There is no evidence that the student has recognised the area of chemistry involved or has given any statement of a relevant chemistry principle. This mark would also be given when the student merely restates the chemistry given in the question.

N.B. It is not necessary to discuss everything to gain a maximum three marks. Remember that the answer does not have to be 'excellent' or 'complete'.

Topic 5: Pharmaceutical chemistry

End of Topic 5 test (page 363)

Q1: 58.5 g of NaCl contains 23 g of sodium.
Therefore 5 g of NaCl contains $(23/58.5) \times 5 = 1.97$ g of Na^+ ions.
% m/v = $(1.97/500) \times 100 = 0.39$%.

Q2: 3.5% of 250 cm^3 = $(3.5/100) \times 250 = 8.75$ cm^3

Q3: 0.75 g toothpaste contains 1.125 mg F^- ions ppm = mg per kg, so 1 kg toothpaste (1000 g) contains $(1.125/0.75) \times 1000 = 1500$ ppm

Q4: 1000 g of toothpaste contains 1500 mg (1.5 g) of F^- ions.
10 g of toothpaste contains 15 mg (0.015 g) of F^- ions.
42 g of NaF contains 19 g of F^- ions.
Therefore, 0.015 g of F^- ions is contained in $(42/19) \times 0.015 = 0.033$ g.
33 mg of sodium fluoride is required.

Q5: D

Q6: c) Hydrolysis

Q7: b) Reduction

Q8: a) $C_9H_8O_4$

Q9: b) Alkene

Q10: a) Primary

Q11: 2

Q12: B

Q13: C

Q14: A

Q15: Antagonist or inhibitor.

Q16: Ethanal has the formula C_2H_4O and the difference between phenacetin and paracetamol is C_2H_4. This means that there is an oxgen atom required for the change.

Q17:

- A tetrahedral carbon atom with four different groups attached to it gives rise to two possible arrangements, each a mirror image of the other. Such isomers are known as enantiomers. They have no centre of symmetry and are described as chiral.
- Since salbutamol functions as an agonist it has to be sufficiently similar to the natural molecule which initiates a biological response to mimic this molecule. The shape of one of the isomers does this well, the other much less well. (one mark for each of the two parts)

Topic 6: Organic chemistry and instrumental analysis test

Organic chemistry and instrumental analysis test (page 368)

Q1: c) sp^2 hybridised carbon atoms and sp^3 hybridised carbon atoms.

Q2: B

Q3: In an isolated carbon atom there are only two unpaired electrons. However, when carbon forms bonds with other atoms the 2s and the 2p orbitals mix to form four degenerate orbitals known as sp^3 hybrid orbitals. These four orbitals each contain one unpaired electron therefore carbon is able to form four bonds.

Q4: A

Q5: C_2H_5OH, ethanol or C_2H_6O

Q6: 2-methylbutane

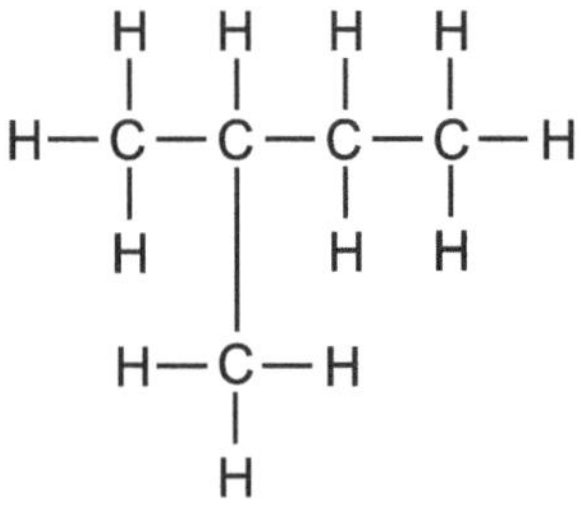

```
    H   H   H   H
    |   |   |   |
H — C — C — C — C — H
    |   |   |   |
    H   |   H   H
        |
      H—C—H
        |
        H
```

Q7: b) Effect on plane of polarised light

Q8: b) $CH_3CH_2CH_2CH(OH)CH_3$

Q9: Four different groups around a carbon atom or mirror images are non-superimposable.

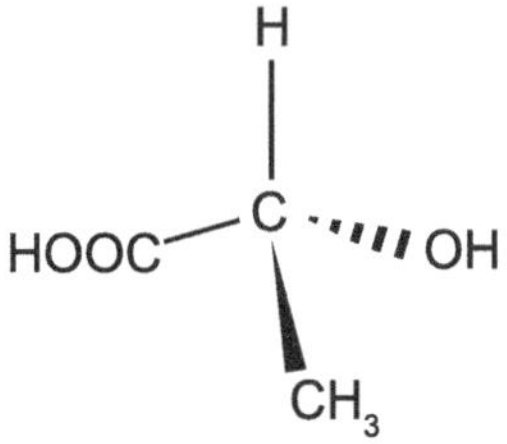

```
         H
         |
HOOC — C ····· OH
         |
        CH_3
```

Q10: b) CH_3OH and CH_3COCl

Q11: b) electrophilic substitution.

Q12: Hydrolysis

Q13: 2-hydroxy-2-methylbutanoic acid

ANSWERS: UNIT 4 TOPIC 6

Q14:

$CH_3-CH\overset{H\frown Br}{=}CH-CH_3 \longrightarrow CH_3-CH\underset{\oplus}{-}CH-CH_3 \quad H \;\; \overset{..}{Br}{}^{\ominus}$

$CH_3-\underset{\oplus}{CH}-\overset{H}{\underset{|}{CH}}-CH_3 \longrightarrow CH_3-CH-\overset{H}{\underset{|}{CH}}-CH_3$
$\overset{\frown}{\underset{\overset{..}{Br}{}^{\ominus}}{}} \qquad\qquad\qquad\qquad Br$

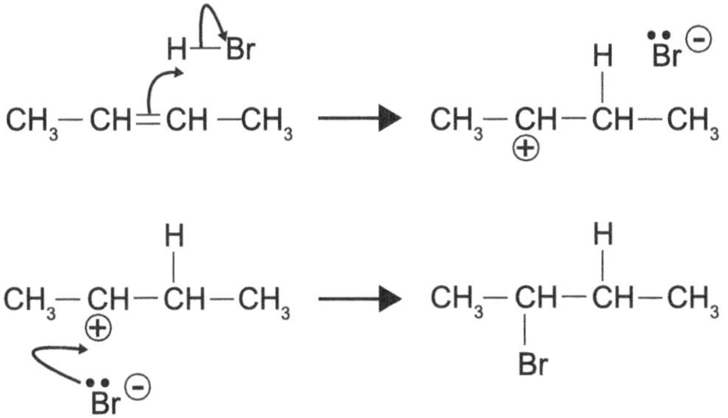

Q15: Aqueous KOH/NaOH or aqueous alkali

Q16: d) $[CH_3CO]^+$

Q17: c) CH_3OH

Q18: a) Infra-red spectroscopy

Q19: Mass of oxygen = 0·1920g

Element	C	H	O
moles	0.0180	0.0359	0.0120
mole ratio	3	6	2
or	1.5	3	1

Empirical formula $C_3H_6O_2$

Q20: It causes them to vibrate.

Q21: b) antagonist.

Q22:

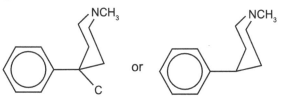 or

Q23: Produces a response like the body's natural active compound/enhances body's natural compound.